Deceived Beyond Belief
The Awakening

PROLOGUE

RENEE PITTMAN

This ebook is licensed for your personal enjoyment only. This ebook may not be re-sold or given away to other people. If you would like to share this book with another person, please purchase an additional copy for each recipient. If you're reading this book and did not purchase it, or it was not purchased for your enjoyment only, then please return to Smashwords.com or your favorite retailer and purchase your own copy. Thank you for respecting the hard work of this author.

Mother's Love Publishing and Enterprises

Copyright © 2018 Renee Pittman

All rights reserved.

ISBN: 978 -1-7374060-5-1

DEDICATION

To the Strongest Among Us

Out of Suffering Have Emerged the Strongest Souls,

The Most Massive Characters are Seared with Scars.

Table of Contents

Preface vii

Chapter One: The Art of High-Tech Persuasion 1

Chapter Two: Prologue 14

Chapter Three: Sex Used as a Weapon Against Us 27

Chapter Four: Id, The Place Where Ego and Libido Meet 39

Chapter Five: Deceived Beyond Belief 73

Chapter Six: Life Long Human Testing 83

Chapter Seven: Eyes Wide Open 95

Chapter Eight: Unification Nationwide and Globally 104

Chapter Nine: Behold, I Send You Out as Sheep 112

Chapter Ten: She Set Herself Free 128

Chapter Eleven: Sheeple vs. Sheeple - Crafted by Wolves 141

Chapter Twelve: Vibrate at the Highest Frequency 155

About the Author 170

PREFACE

The North Wind and the Sun

The North Wind boasted of great strength. The Sun argued that there is great power in gentleness. "We shall have a contest," said the Sun. Far below, a man traveled a winding road. He was wearing a warm winter coat. "As a test of strength," said the Sun, "Let us see which of us can take the coat off of that man."

"It will be quite simple for me to force him to remove his coat," bragged the Wind. The Wind blew so hard, the birds clung to the trees. The world was filled with dust and leaves. But the harder the wind blew down the road, the tighter the shivering man clung to his coat.

Then, the Sun came out from behind a cloud. Sun warmed the air and the frosty ground. The man on the road unbuttoned his coat. The sun grew slowly brighter and brighter. Soon the man felt so hot, he took off his coat and sat down in a shady spot.

"How did you do that?" said the Wind.
"It was easy," said the Sun,"
I lit the day. Through gentleness I got my way."

~ Aesop's Fable

YOUR GREATNESS IS NOT WHAT YOU HAVE, IT'S WHAT YOU GIVE

NOTE: This book was written under duress, Electronic Surveillance, Military COINTELPRO, beam subjugation Active Denial System subjugation, etc., which includes relentless efforts to destroy the information presented to the public, for obvious, good reason. The truth is powerful!

The question is "Can a Computer be Hacked if it's Not Connected to the Internet?"

An article, with this same title, dated March 23, 2017, by Michael Guta in "Technology Trends" reports it can, reporting:

"A New York Times article reported about NSA technology allowing hackers to get into a computer, even if it is not connected and alter the data. But this technology requires physical access to the computer. According to the Times report, "In most cases, the radio frequency hardware must be physically inserted by someone, a manufacturer, or an unwitting user."

This however is not the only way unconnected computers or smartphones can be accessed or monitored. An article in Business Insider reveals several ways in which this can be achieved. This includes electromagnetic radiation spying, power consumption analysis, using a smartphone's accelerometer as a Key Logger, radio waves that intercept the most secure of networks, using the heat generated by our computer, and accessing data through steel walls.

It must be understood that the author of this book series, has a high standard in all endeavors, and continues to make great effort to bring forth quality work.

CHAPTER ONE

The Art of High-Tech Persuasion

Disguised as new technology, Fox News reported that the US military has begun testing Artificial Intelligence brain implants that can change a person's mood. These mind control chips emit electromagnetic pulses that alter brain chemistry by deep brain stimulation. The technology delivers high-frequency electromagnetic stimulation to a targeted area of the brain. The stimulation changes the electrical signals of the brain thereby changing mood and behavior.

In reality, this so-called 'new' technology is part of continued advancements by the Defense Advanced Research Project Agency (DARPA) a branch of the Department of Defense, which develops new technologies for military use.

Researchers from the University of California and Massachusetts General Hospital report that the artificial intelligence algorithm detects brainwave patterns in the brain specifically associated with moods.

Through high-tech monitoring and tracking of a person, implanted with electro lodes, for example, for seizure disorder monitoring over the course of three weeks, researchers were able to create an algorithm to decode the test subject's moods. The study included images of numbers and identifying emotions on the face, difficulties in concentration and problems with empathy. The research resulted in a greater understanding of the full range of mood disorders and how they can be manipulated. Although the chips won't be able to read a person's mind, the reported stated, the Artificial Intelligence chips do raise ethical concerns. Why? This is because the research has access to activity that encodes human feelings.

The question is what are the moral implications of any technology that has access to activity that encodes feelings?

The full findings of the study related to Artificial Intelligence implants and mood management and behavior modification are published in the journal Nature News.

The image below gives an example of the process, followed by one of several official patents.

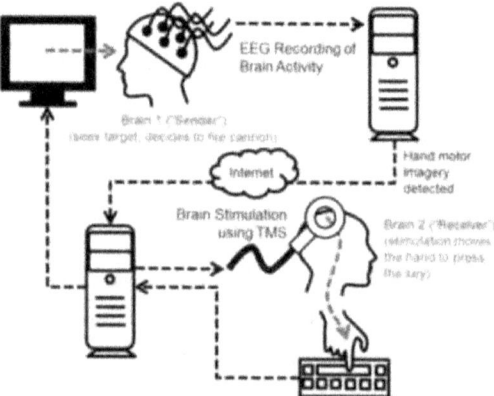

Device for the induction of specific brain wave patterns

Abstract

Brain wave patterns associated with relaxed and meditative states in a subject are gradually induced without deleterious chemical or neurological side effects. A white noise generator (11) has the spectral noise density of its output signal modulated in a manner similar to the brain wave patterns by a switching transistor (18) within a spectrum modulator (12). The modulated white noise signal is amplified by output amplifier (13) and converted to an audio signal by acoustic transducer (14). Ramp generator (16) gradually increases the voltage received by and resultant output frequency of voltage controlled oscillator (17) whereby switching transistor (18) periodically shunts the high frequency components of the white noise signal to ground.

Inventors: **Williamson; John D.** (North Canton, OH)
Assignee: **Omnitronics Research Corporation** (Akron, OH)
Appl. No.: **112537**
Filed: **January 16, 1980**

The ability to alter our mood in actuality removes the intimacy, authenticity and immediacy of our emotions becoming synthetic feelings technologically. And, when does altering brainwave functions and brainwave patterns for positive or negative moods become mind control?

All of our thoughts, sensations and actions arise from bioelectricity generated by neurons in the brain which are then transmitted through complex neural circuits inside our skull. Electrical signals between neurons generate electric fields that radiate from brain tissues as electrical waves that can be picked up, copied and stored into a Brain Computer Interface (BCI) as your unique biometric signature. And, contrary to this type of technology being publicized new, it has been the focus of the Military Industrial Complex for decades.

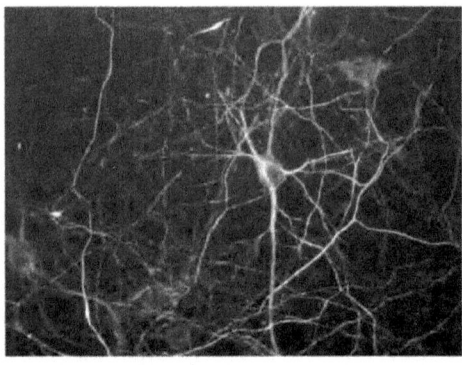

So fundamental are brainwaves to the internal working of our minds, that they have become the ultimate legal definition drawing the line between life and death.

A report by the "International Journal of Research" (IJR), shown in the image above, can be located online and gives an in-depth review of advanced brainwave technology in full use today.

The importance of brainwaves research and influence lies in the capability for alteration and the ability to change a healthy person's conscious and unconscious mental activity. Manipulation of brainwaves can also include states of sexual arousal and the capability of arousal of other emotions including joy, sorrow, stress or agitation.

Today these are patented capabilities of advanced technology for mood modification and thought systems shown in just one Abstract below.

Would these technological advancements be used to create negative moods and extreme emotions, detrimentally, that is so intense that a person acts out from the manipulative beamed influence? Can the technological capability to clone moods result in the creation of detrimental behavior, murder, and ultimately synthetic self-sabotage of a life by those at the helm of advanced systems and devices?

Could the technology activate depression so deep and unbearable that a person finds that just waking each morning is a great burden under the intensity? Are Federal, state and local police, operating from highly advanced operation centers today using these advancements? And if so, could a stubborn target that refuses to bend or break be nudged to commit suicide? The answer is yes, yes and yes.

Since the human nervous system registers a response milli-seconds before an action is carried out, it is today possible to capture the signal and translate it to a corresponding action immediately. And, the technology is becoming mainstream. For example, Facebook reported it wants to transcribe thoughts to speech, and Nissan is talking of controlling cars using thought and brainwaves.

By 2018, the high-level joint targeting operation around me encompassed USAF personnel, the Los Angeles Police Department, both overseen and spearheaded from the FBI, Joint Resource Intelligence Center in Norwalk, California. And I can tell you from personal experience that yes, this technology can destroy a life by mood modification, if the focus is unaware of the capability.

Hoping to destroy me, I was being hit hard from drones assigned to track and monitor me equipped with psychotronic weapons. Using the mood-altering intense frequency beam, the relentless hope was the creation of intense negative feelings that would force me to act out negatively or self-destruct. The battle was for my mind and very soul. It began every morning as soon as I opened my eyes. For impact, monitored in real-time around the clock, the attacks were always prefaced by negative verbal insults through the drone communication system, Parametric Speaker or another type, the Hypersonic Sound System.

> **Thought controlled system**
>
> Abstract
>
> A system for controlling a computer by thoughts in the user's brain. The system applies stimuli of the brain via magnetic source imaging (MSI) for controlling the computer. Patterns of brain stimuli are recorded along with the particular thoughts that produced them and these thoughts are interpreted as functions for controlling the computer in much the same way as inputs from a keyboard or mouse. Criteria of acceptance of thought stimuli are generated by the system. Body stimuli, in addition to the brain stimuli, are monitored and used by the system. A user profile is maintained and displayed along with selected pictures for assisting with stimuli/thought pattern utilization. Artificial Intelligence is used to enhance stimuli selection, human factors and reliability, as well as analyzing past errors, adverse occurrences and performance. Analyses and summaries are produced by the system for psychiatrists, psychologists, researchers and users to study system enhancement, biofeedback, psychological impact, brain activity, localization and identification of feelings and thought patterns. Stimuli are monitored at brain and body locations. Various functions are applied to animals.

The targeting was reinforced by thought monitoring around the clock. And any negative passing thought deciphered by the high-tech mind reading capability, could, if I were not aware, become my worst enemy and used strategically against me. One thing is certain, this effort wants me gone, silenced or dead. The stage had been set when silly me, who should have known better, after twelve years of PsyOps tactics, a few months before finalization, monitored every step of the way, went to the VA hospital. My reporting that the "FBI is after me" resulted in, what appeared to be a strategic useful diagnosis place into my health records, then giving something beneficial for the targeting operation to work with to silence me.

The crazy tag continues to be one of the most powerful discrediting methods applied to many which have stood the test of time for these efforts silencing for decades.

Whenever any negative emotion materialized or I was engulfed with unexplainable great sadness, sorrow or regret, out of the blue, beamed back to me, awareness would be the key for life-saving. Trust me when I say it was intense.

In fact, shocked at what is happening today and for years, my heart broke for those, who unfortunately, had not the remotest clue that they were in this ongoing nonconsensual human experimentation program. In a childhood notion, I declared as a little girl that I wanted to help save lives one day. Was this my opportunity I think now, amused, in remembrance.

In reality, what choice did I really have? I am under the gun by focus 24/7. Far be it for me to be afraid to fight for my life. I embarked on a campaign of exposure up against those who probably thought I was from the Ape family and this perception resulted in my not being considered human and because of this expendable.

Ultimately, it was awareness of the patented capabilities used by these joint targeting efforts that curtailed their success around me. Awareness is how I now believed I could help others. I knew that awareness had continued to save my life.

With me, to counter the affect, I first started to laugh. I laughed at the absurdity of what the effort was trying to accomplish. The depression was thick as it slowly engulfed me sporadically throughout the day although I was in a relatively good mood. I knew that the result of laughter and amusement was two-fold, one, it instantly changed my mood, releasing refreshing dopamine and two, conceiving what they were doing as amusing, saved my sanity because I did not take it seriously. On some of the heavier day attempts, to literally bring me down, I took a break from this manuscript and went to the gym. Exercising to combat the frequency manipulation also released dopamine.

Adam Piore explores how bioengineers are harnessing the latest technologies to unlock untapped abilities in the human body and mind.

This includes translating neural brain patterns of thoughts into words. In reality, these are old technologies being reported anew.

Menticide is described as a systematic effort to undermine and destroy a person's values and beliefs. The goal of menticide is to induce radically different views and ideas against the true self or true nature of a person.

Menticide is accomplished by systematically changing a person's responses to life and by successfully rearranging, thoughts, emotions, and feelings. The result is the development of personality characteristics and traits that are injected from an outside source.

Menticide can be achieved through prolonged interrogation, trauma-based dissolution of the mind, then broken and compartmentalized, and by the prolonged psychological warfare this effort hoped.

Many targeted today are those who have been used and abused as human experimentation subjects for decades, with the program unleashed after official legalization for specific agency testing, post 9/11. These agencies are specifically the military and law enforcement. The result is the evolving a high-tech, monstrous, technological, technocratic state, nationwide and reported globally.

In this new paradigm, working hand-in-hand, is, and has always been, strategic mind weakening drug synchronization. History reveals drugs have consistently been part of the global agenda. With this understanding, you begin gaining a full understanding of how "The Program" is structured.

It is not by chance that there is an epidemic of Heroin addiction in not only poor communities, but also affluent America and spreading across the country like wildfire, and drugs so powerfully addicting most do not stand a chance. These are the aspects of the strategic, mass, social and population control mechanisms which fall in place. With opening eyes, you then realize there is a connection with 90% coming from occupied Afghanistan. You then take a good look at who benefits. And when you do, you awakened to a diabolical scheme.

Years of ongoing research into mass population control techniques, long recognized that the human mind can be reshaped and remolded into a "Herd Mentality" when weakened or chemically altered. And when looking at the full picture, with efforts backed by mass media strategic distractions, the intent is non-focus on reality. These are prime example of programming designed to create our reality for us by the outside sources that be. Add to this patented, technological psychophysical, systems and devices which can be used for behavior and mood management of the nervous system and voila! George Orwell's fictional omen comes to light when he wrote:

"Your worse enemy, he reflected, was your nervous system. At any moment the tension inside you was liable to translate itself into some visible symptom."

George Orwell "1984"
And, it is capable on a large scale.

Hendricus Loos is one of many inventors, whom I mention often, as a forerunner in the global control Matrix patents and an arsenal of weapons today designed to focus on the nervous system.

And, these bio-weapons and their influence are unpreventable. This is due to all biological life resonating to the same frequency, revealed by the Schumann Resonance, and the Divine Order of this planet. When you think of the magnitude of this, you begin to understand the true oneness of humanity, which occurs each night at a specific time. You begin to understand a bioelectric powerful truth. Those developing technology, for ill, understand this as well.

The brilliance of the influence and frequency control is that it is not obvious. This is the understatement in Aesop's Fable and the challenge between the Sun and the Wind focused on the oblivious traveler and his unawareness of his strings being pulled. It is a method of influence that operates in vast space and time, within the Electromagnetic Spectrum by harnessed energy, honed for spiritual wickedness. It is fueled not by a vibration of a higher frequency, love being the highest, or any act of love, created from positive thought, word, deed, action, but used, monstrously by global "Beast System" and those involved worshippers of the Beast.

The ideation and hunger for global control, death, destruction and bloodshed, its fuel, and the machination and desires to do so, argumentatively originating from the bowels of the metaphoric Hell.

How also can one explain psychological electronic attacks, and focused destruction on the essence and the very soul and spiritual path of human nature, and more importantly, not only men and women but children.

When consciousness is intentionally changed by redirection into something a human life was not meant to become, without a doubt evil power and control are the motivating factors. In actuality, this dynamic is similar to the biblical account of a birthright stolen told in the story of Isaac and his sons brother's Esau and Jacob.

"He took my birthright, and look, now he has taken my blessing" says Jacob. Jacob then asked, his father, "Haven't you saved a blessing for me?"

The fact is, I have learned and hold dear, an understanding that through the toughest of storms, God wants to give us the ultimate blessing that we desperately need and seek, resulting from blind faith and trust. I also had to develop an understanding, if I were to survive, that God uses human deception to accomplish the redemption plan and to ultimately bless the world from darkness to Light.

In Book I through Book V of this book series, I set the stage for the official targeting program, by documentation of specific agencies involved historically, and through my personal education based on experiences. However, in hindsight, in the midst, I realized that my education began a long time ago and a feeling that something just did not feel right. It began long before I, and many other knew this program exists or that we were in it.

Today, twelve years later, with no sign of relief until I am dead, the covert, ongoing hope, to silence not only me but many, continues in the form of high-level agency plausible denial tactics, insinuations of my sanity, and hope it will stick, and ongoing harassment by covert technologies.

Through the storm an awakening evolved through connections that could not be denied. As a small light began to flicker than glow, admittedly, it was Earth shattering.

How do I know so much about this program? It is because I have lived it through identical experiences, dating back to age four or five today in full recognition. Spurts of memory of identical specific experiences both today and yesterday became the connection.

It would take time to formulate everything and some memories were too painful to reveal so here I detail less than 50%. However, one thing is certain, the revelations pointed to the likelihood of an outside scientific source operating literally under the radar. In the grand scheme, it makes complete sense that a program like this would exist and continue to flourish for decades. How else can a small few, the 1% control 99% of the population?

Jon Rappoport nailed it when he wrote an article titled:

"Mind control: The Pentagon Mission to Program the Brain"

"Since the dawn of time, the most powerful groups in every society have practiced forms of mind control on populations. They determined it was necessary.

Eventually, they decided it was their most important job. Convincing the masses that a fabricated reality is Reality...that task requires formidable mind control."

Today DARPA, and the Department of Defense are leading the way on a mission to program the human brain anew and widespread.

This book is based on epiphanies resulting from flashes of light, demanding that I repeatedly travelled back in time, down memory lane, in the search for how and more importantly, why?

As enlightenment took take shape, fueled by the courage to dig deep into confusion and the pain caused by confusion, I debated if I should share some of these experiences at all. Based on today's DSM-5 it could play into the hands of operatives of this program seeking to discredit me as a wackadoodle. There had already been three attempts to lock me up in psych wards and throw away the key. If so, another troublemaker would disappear, but each time the effort failed.

I thanked God for a clear eye and set out creating this manuscript albeit under the circumstances cautioned. This was difficult for me because I am an upfront, straight shooter. I remembered warnings from my surrogate Grandmother, "never give a person a stick to hit you over the head with." However, I had to trust that the truth will prevail, backed as well by too numerous to mention confirmation of similarities told to me by others experiencing the same plight and our unified confirmations gaining momentum.

In the global awakening today, in the "Age of Information," I and many others have begun to understand that it is not out of the question that we have been targeted in a menticide program since early childhood. In fact, if we understand that Project MKULTRA never really ended, but went underground and continued to flourish. What is happening today is simply, a hideous extenuation of the progression of a never-ending agenda.

Deceived Beyond Belief the Awakening

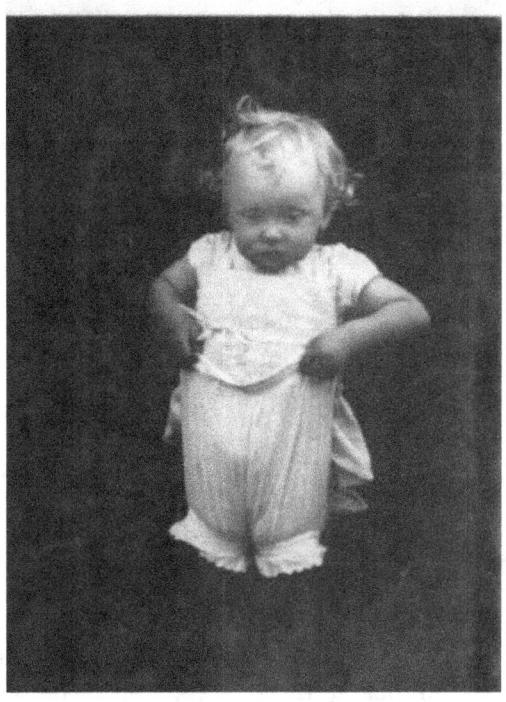

So, I pull up my big girl britches and carried on...

CHAPTER TWO

Prologue

It may appear that I am dwelling on Sex in this book, the good, the bad, and the ugly. It is true, too some degree I am. This is because sexual energy is, and has always been, used as a major technique in mind control programming, deviously. This is especially true in goal orientated trauma-based mind programming. Sex has been used successfully, because of awareness of it having the powerful energy and the capability to divide the human mind.

With me, it would be the exact type of sexual stimulation, I experienced as a child, combined with influence technologies, connecting todays hopeful subliminal influence control testing, that became the key to unlocking the squeaky door. The difference being today my complete awareness of official targeting and the knowledge gained from the experiences of patented technology, and again, also confirmed as identical by numerous reports of many other targets.

When the possibly of a connection materialized, it hit me like a ton of bricks. The possibility hurt deeply. This was because if true, the focus was on tampering with the innocence of childhood by a high-level scientific program that apparently believed the innocence of a child perfect for use as human guinea pigs. The connection made the confusion of unwanted sexual stimulation, today, and as a child, now logical. It absolutely could not be a coincidence.

Beginning in 2006, in the early stages of the ongoing nonconsensual human experimentation effort around me, the men in this program started toying with me sexually using the sexual stimulation patent of Loos. It was extreme. Some nights, while sleeping, I was brought to a full orgasm, technologically, and against my will. I remember thinking do women have wet dreams and first though apparently so. The nerve pulsating in my private area had accomplished their goal.

The beamed affect never had any type of mental motivation from me. It was never stimulated by arousing thoughts, mental desire or attraction and purely physical. I questioned why because of this and also wondered if something was wrong me to the point of my not understanding myself which is scary. Clue was that these experiences lacked an intellectual connect.

If I was not born with a God given, willful spirit, and obvious inquiring mind, I could have been effortlessly changed into something I am not, or meant to be, as the wind storm blew in the form of a known scientific challenge trying and trying and trying to change the very essence of me.

The main issue became the recollection of the intense sexual stimulation around little girls while very young. Getting caught and spanked while playing house on the back porch with pants down and two little girls rubbing against each other for an orgasm also made an imprint of the experience. How could a child so young be brought to a full orgasm at such a young innocent age? I had often thought and wondered throughout the years? If I were in fact part of ongoing human experimentation testing, it did not the answer was clear. However, it did not create acceptance but instead frightened me, because it was girls, and the fear lingered all of my life.

Today unwanted technological sexual stimulation of both women and men, and likely also children, is probably continuing and could play a major role, if so, in pedophilia and human trafficking motivation which is widespread. The technology has been around for a long time, and it took many test subjects to bring it to perfection before patenting.

We can look back, at least officially, to the 50s after Nazi scientist were brought from Germany to the US after World War II to continue their work in mind control as a system of mass population control.

In the early stages of the relentless, now official psychophysical effort around me, the beamed urge to masturbate was so intense while alone in my room materializing out of the blue anew in 2006. The sensation would engulf me demanding that I had no other choice but to try and relieve myself. This was by the men in the operation center and I know done for their amusement. Today, it makes perfect sense that seventy percent of those targeted are, single women, living alone and that ninety percent of those at the helm of this technology are apparently deviant men, creating operation center real-time reality show entertainment. How bored they must become from the surveillance during eight-hour shifts while watching people lead a typical normal life, and boring activity. They must want to spice it up a bit.

Several of the patents by Hendricus Loos, defined in the Abstract, describe the ability for sexual stimulation and physical sexual arousal. And, major testing unfolded in many areas decades ago reaching high points during the 60s, 70s, 80s and 90s.

I have never had any emotional sexual attraction or desire for women, although having close best friends throughout my life. And, the childhood experiences, happening precisely three times, still lingered as confusion. What stood out in my mind was that the sensation materialized out of nowhere, which I recall to this day. I questioned for years was it simply nothing more than child's play?

As a result of trying to make sense of it all, and in awareness today of the patented technological ability, and for the rest of my life being not inclined towards girls, I concluded it possible that homosexuals, reporting they were Gay as children, could have possibly been nonconsensual human guinea pigs, in a massive Frankenstein scientist program hoping to change sexual identity.

Without a doubt there has always been a historical, scientific focus, on behavior modification. Changing people is essentially what behavior modification is all about.

This awareness, and the connection of identical experiences today, and as a child, also freed me from the fear and turmoil that because of those early experiences, that I was a possibly a closet Gay. The awareness also cured me of the intense homophobia I harbored for years afterwards scared that I was. There was now a method to the madness, and an explanation revealed by those at the helm today. With the exception of the few childhood experiences, and the insatiable sexual urge to rub against another little girl and the full orgasm at such a young age, I now had a clue that demanded research for further confirmation and speaking with other targets.

There was my friend in Canada who reported how he did not understand the intense sexual stimulation around his mother as a child and another reported the sexual arousal that kept him sexually aroused for the most part and the belief that he was a stud.

The fact is Ewen Cameron spearheaded the program in Canada also beginning in the 50s connected with United States interest and the Association of Psychiatry. Another confessed his belief after his eye opening of being in this program, that he likely had been made into a homosexual and now also had HIV. Another reported that for years, posted on Facebook, there had been beamed influence focused on him for pedophilia and stimulated also around young children as an adult.

The fact is homosexuality dates back eons with Athens, Greece being a hot spot.

One mythological tale originates in Ancient Egypt describing the extended conflict between Osiris and Seth depicted in the biblical tale of Cain and Able. Seth, the rival brother, murders Osiris then seeks to remove Horus, Osiris' son and heir, from the throne with Seth's claim to be king of the gods.

This narrative, is referred to as "The Contending's of Horus and Seth," and told in different versions as far back to the early Middle Kingdom (2040-1674 B.C.). There are also hints of homosexuality in the "Egyptian Book of the Dead."

Was there a technological connection even then?

One thing is certain, awareness of the Electromagnetic Spectrum dates back to the beginning of time. In fact, the Tower of Giza was empowered by the electromagnetic energy from the Spectrum. If so, knowledge of toying with bioelectric human lives is antediluvian.

One version, of the Egyptian tale, describes that Seth said to Horus: 'Come let us spend a pleasant hour at my house.' Horus answered, 'With pleasure, with pleasure.'

When it was evening a bed was spread for them and they lay down. During the night Seth made his penis stiff and he placed it between the loins of Horus. Horus put his hands between his loins and caught the sperm of Seth. Then Horus went to his mother, Isis, and said: 'Help me...! Come, see what Seth has done to me.' And he opened his hand and let her see Seth's semen. With a scream she took her weapon and cut off his hand and threw it in the water and conjured up for him a hand to make up for it." Apparently, she feared the possibility of the influence.

It must be understood. I can't speak for others and the personal experiences in their lives but only my own. Today, I now firmly believe that there was an outside source, attempting to motivate me, based on what was happening, then and what I know as factual strategies today. Setting a person up for blackmail after influenced into acting on some type of sexual deviance, against their nature and character, you must admit, could also be a part of an effective control mechanism or used for destruction.

Admittedly, again, true also, some would call these types of childhood experiments as innocent child's play. However, the experiences haunted me throughout my life as an anomaly to my true self. This was not because of the interaction while playing house with little girls and later little boys after we moved, but again, the intense sexual stimulation and more importantly, again, a full orgasm at age four through age six.

Even more frightening to me was the psychological concept that it is abnormal for people to be Homophobes. Some psychiatrist believes that one should look deeply within self for a connection to the fear. And reportedly, their studies found that those most hostile toward gays or hold strong anti-gay views may themselves have same-sex desires. This was a troubling concept yet again. Essentially the study sought to deny that no genuine mental attraction to the same sex holds any validity, and that any rightful fear is in fact opposition to the real self, and an unknown part of self.

My conclusion, rubbish!

However, I must admit that willingly harming Gay people based on sexual preference or worse indeed carries some type of deep-seated psychosis resulting likely from the factual intense programming taught in religions.

The problem here is that I never was hostile, or held any, hatred or animosity towards any human being nor for any reason. I was just scared to be around gay people thinking I could become Gay or that it was a Spirit, based on the unwanted stimulation I experience which could attach itself to me by bringing it into my environment by association. I stayed clear.

It must be understood that during the 60s, we lived in an era where Lucy and Ricky Ricardo slept in separate beds, and unlike today it was taboo during this timeframe to even share kiss on national television. Sex-scenes were once verboten for any hour of network programming unlike today.

Today, sexual scenes or reference to sex are relentlessly exploited everywhere you turn creating subliminal influence without a doubt. This includes selling sexual arousal and sexual enhancing or stimulating products during primetime hours by commercials and happening while children are watching.

I was raised by a grandmotherly figure, and she was in her seventies, single, separated from her husband, and it just her and I as a child growing up. I never saw anyone having sex around me, nor did I watch the old black and white television at all.

One of the objectives of mind control is too subtly, influence subjects over a period of many years. A person is subtly nudged into thoughts of various types of activity, although many do not act. The windstorm creates, in some cases, severe emotional strife, which then has the powerful capability to lower self-esteem and mold the subject. The result of confusion is that a person can become pliable to both overt and covert influence.

Self-doubt is powerful and impacts mental stability and especially when combined with resulting self-loathing. The result is that a subject or subjects become programmable by the outside source, controlling physical reactions, sensations and thought. The subject then loses his or her identity, if this makes sense to you. They believe that suggestions or sensations are actually their own, becoming being intensely conflicted and accept whatever the programmers seek.

As I looked back, I first needed to understand the turbulent times prevalent in the 60s. It was a time of the free love, the sexual explosion and mass trance of the Hippie movement fueled by hallucinating drugs of the time used with the specific goal for programming testing. It was a time when opposition to the Vietnam War had successfully mobilized Vietnam protest, by many famous public figures. The 60s was an era of assassinations, riots, and the emergence of programmed Manchurians believed under hypnotic influence and used as plausible denial puppets for Intel agencies. There were many unaccountable covert, successful, silencing efforts of many high-profile figures of the 60s, the Kennedys, Martin Luther King, John Lennon, and others.

In California, specific agencies, such as Health and Human Services, specific universities such as Stanford, UCLA and USC, and Loma

Linda Hospital, to include the Loma Linda VA hospital in California, revealed by Freedom of Information Act information documented the focus of behavioral modification programs funded through government allotted dollars for that timeframe.

The most dangerous hospital in America, reportedly, with funded research is near the University of Michigan. It is the VA hospital in Ann Arbor, Michigan. It reported was a lab focused on Remote Neural Monitoring (RNM) and decades of mind control experiments reported by victims as crimes against humanity. In fact, the first reports of Remote Neural Monitoring attacks surfaced throughout Michigan bordering Canada and many Canadian border communities.

Extremely Low Frequency radio signals can pinpoint a target anywhere on Earth and can penetrate water, rock, concrete and other dense matter. There have been many studies focused on frequencies as far back as the early 1900s by Nikola Tesla.

The brain is composed of neurons (wires) and is powered by low electrical currents. This is how insects communicate with non-contact antenna and humans can communicate with radio transceivers.

Our biometric signature has a unique genetic code (DNA) and each of us has our own distinct radio frequencies. If you find the frequencies terror technician use, that can communicate directly with the brain, nervous system and body, it becomes a formidable battle of wits for the targeted subject. A staff of technicians could easily isolate a victim and beam radio signals into the victim's home as well happening today all over the world.

In addition to mind control torture, a methodology to produce remotely-delivered body shocks to the groin and elsewhere also evolved. This yet again, reveals ongoing, intent focused on and the awareness of the controlling aspect by focus on genitalia to include using pain for submission.

Interestingly, Michigan was one of the first states in the nation to enact legislation outlawing electromagnetic harassment. The enactments, lends credibility to the fact that these covert operations exist are used.

Over the years, race has never been the defining marker for who would be targeted, although the black community would be heavily hit. The oppression by design of the black race is clear throughout history by disenfranchisement, massive intentional unemployment, etc., resulting in institutionalized racism. There also appears to be a decisive focused effort to keep wealth, black ownership of business to include homeownership out of the black community, which passes on generational wealth. The goal is to keep the black community impoverished or perceived as, ignorant, shiftless, incapable and lazy.

For example, the Tulsa, Oklahoma race riot, also known as the Tulsa Massacre, took place from May 1921 to June 1, 1921. A white mob attacked and destroyed prosperous members of the black community known as the "Black Wall Street" in the city that was thriving.

There has always been the Nazi ideation and Nazi agenda at the foundation of this program, to include the Nazi useful and clever connection to Nazism and Psychiatry. And, without a doubt, one thing is certain, the underlying foundation for mental control continues to be to perpetuate a culture of "ism" be it racism, sexism, and classism.

In the early 1950s, some very odd experiments were being performed at Tulane University in Louisiana. Dr. Robert G. Heath found that he could manipulate the pleasure and pain centers of the brain by surgically placing electrodes deep inside.

Septal stimulation was the constant of Dr. Robert G. Heath's career, however, he engaged in an enormous variety of other work, publishing at least 425 papers. Among these were his efforts to treat gay men by turning "repugnant feelings ... toward the opposite sex" into pleasurable ones – and similar work on "frigid women."

Dr. Robert Heath's work on mind-control at Tulane was reportedly partly funded by the US military and the Central Intelligence Agency.

Dr. Heath's subjects were African Americans. In the words of Heath's collaborator Australian psychiatrist Harry Bailey, this was "because they were everywhere and cheap experimental animals"

The patients would be wired up and given a little box and "just went around, 'pop, pop, pop', all the time, resulting in continuous orgasms." Following the discovery by Olds and Milner of the "pleasure centres" of the brain [James Olds and Peter Milner, "Positive Reinforcement Produced by Electrical Stimulation of the Septal Area and Other Regions of the Rat Brain," Journal of Comparative and Physiological Psychology 47 (1954): 419-28.],

Later, Dr. Heath admitted in print that septal stimulation had different effects on different people – generally serving to amplify rather than create emotions, especially in the case of arousal, and having much less effect on those who were already feeling happy and contented.

In fact, a woman called Claudia Mullen even testified before Congress in 1995 that Heath had, when she came to him as a child patient, engaged in all kinds of unethical practices before handing her over to the custody of the CIA, where she was used as a sex slave. He has been accused of mind control, of barbarity, of "Nazi science," which included using prisoners in Charity, Jackson and elsewhere as his playthings.

Before his death in 1999 Dr. Heath was known, not as the man who was the first to map out the pleasure circuit of humans, but as a man whose work seemed closer to monstrous science fiction than practical, legitimate humane or moral medicine.

Many ask, "Where does the Irrational Hatred of Blacks Come From?"

Kevin Rogers, Quora writes:

The hate isn't entirely irrational and that almost everything is rational if you understand how it came about.

The United States became extremely wealthy largely because of slave labor. When slavery ended, after the Civil War, promises made to ex-slaves were often broken. Former slave masters knew they weren't being just, but they were angry that they had lost an easy path to prosperity. The fear at the time was that black people in sheer numbers could turn violent and kill the wealthy white slave owners out of revenge.

How do you overcome that? You reduce or eliminate the rights they are supposed to have in order to reduce the ease in which a rebellion can be facilitated.

How did this happen:

Make it more difficult for black people to vote.

- Segregation, legal or otherwise, which still goes on today to a lesser degree. Schools in typically black neighborhoods are underfunded because of the way schools are funded.

- Discredit black social movements that are bringing about real social change (Black Panthers, BLM).

- Make financing more difficult blacks thereby making things like home ownership and entrepreneurship a greater challenge for the black population.

- Portray negative stereotypes in the media. People from areas where not many blacks live will have only these images to base their opinions of black people on. Blacks meeting people from these areas will have to work harder to overcome stereotypes.

    These are processes that have hinder black people from advancing to their full potential in the United States.

    In the process, it has given opponents of black progress purely psychotic ego driven and believed strong proof that blacks are inferior. For example,

- If blacks cared, they would come out and vote. • Blacks are less likely to own homes. Blacks are more likely to be in prison.
- Blacks always play the race card.
- The media shows the truth (did you see the facts above?)

All of this is without a doubt still going into some of the more heinous examples of actions against blacks in America even today.

Here are some articles; Kevin Rogers' suggesting just in case you are wondering what he is referring to.

Bringing crack cocaine into black neighborhoods - Key Figures in CIA-Crack Cocaine Scandal Begin to Come Forward

Intentionally infecting black males with syphilis Tuskegee Study - Timeline - CDC - NCHHSTP

Now, if you had taken part in any of these things, wouldn't you be afraid of blacks, too says Rogers? Of course, so you continue employing the same tactics in slightly different ways, in order to keep things as they are which of course only increases your fear.

Slavery ended 150 years ago. Racism is still going strong.

People fear blacks because they don't want to look at the reasons behind the reasons. Racial issues are a complicated. There are many ways to distort the facts.

Admitting that these things go on would be admitting that things like affirmative action are not only fair, but just.

Admittedly, personally, in full awareness of the control mechanisms in place, personally, I wondered if programming of wealthy athletes, and entertainers, both segments of intense conspiracies of mind-controlled, control of the entertainment industry, would influence them towards white women, done while denigrating black women exploited in rap music used also to keep wealth from trickling down to the black community resulting in growth and development. Yes, it does sound like Conspiracy Theory, however, you must admit, it would be a very clever method of oppression insuring that money that is redirected to other communities does not strengthen the black community. When the athletes marry white, the money is returned to the white community and ultimately generationally.

The black community in the 60s specifically after the Watts Riot, likely became a major focus for mind control and behavior modification experiments focus because of the rebellion.

The result of the uproar was more than likely, a massive scientific focus.

CHAPTER THREE

Sex Used as a Weapon Against Us

In 2014, seeking to add to the Preface, my encounter with Robert F. Kennedy at the age of seven in Book II, "You Are Not My Big Brother – Menticide..." detailing how the targeting effort around me became official and to light in again 2006, there were no media photographs or coverage of the Senator's motorcade's brief detour through a predominantly black community in Los Angeles. I began a search for people around me at the time, knowing I would need confirmation. When a now deceased person asked me years later, did I remember the day, as she put it, that "Robert Kennedy came and got me" I thought the addition ultimately mirrored a destined path for me. Robert Kennedy was assassinated by someone who had all of the characteristics of a Manchurian.

One of the children, I played with as a child was named Richard. He was a puny little thing, very sweet and very sensitive. He was also relentlessly teased mercilessly by the other boys as being a Sissy the old title for Gay. We made contact and decided to have lunch. One thing mentioned over the phone before the lunch date, which was heartwarming was his gratitude for me that I was the only child consistently kind to him during those traumatic years for him with all of us under ten years of age. I had lost contact with these friends over many years.

Now at lunch at Red Lobster, Richard told me that he had married a very pretty young woman, in his twenties, in the 80s, had a son and now was one year divorced. He also told me that he had recently come out as Gay.

Because I had a hunch that something else was afoot again while we were young, because of my childhood sexual stimulation connection, I was determined to explore the possibility and hope for any confirmation by another around during that time. During lunch, I asked Richard point blank, after we settled in and he devoured lobster, crab, and corn on the cob, did he remember being sexually stimulated as a child? There was an instant look of recognition in his eyes, and he was stunned at my question. He replied yes, without hesitation. I asked him did he recall being sexually stimulated around little girls and little boys. He replied around little boys.

Now 2014, during this conversation, he reported that after struggling for years with his sexual identity beginning in childhood he woke up one and decided, why fight the feeling, that he was Gay. He divorced his wife who adored him, and now in his 50s began to live a serious Gay lifestyle. He told me he was convinced that he was always meant to be gay and that it his destiny.

Could it have been subliminal influence combined with him admitting sexual stimulation as a child I wondered? One of the other people, now an adult, immediately went from zero to one hundred, becoming highly agitated when I tried to question him. In his case, and that of many, he had been taken through the intentional drug infiltration craze of this community in the early 80s which had begun taking shape, shortly before when I went into the military. As a result, any potential he or others had, almost the entire population, had been successfully destroyed. I recall he was extremely bright with great potential, although today he is barely getting by. He became threatening when I asked him. However, his mother and younger brother had had success by working, mother for law enforcement, specifically LAPD, and other son as a probation office.

When the mother stopped taking my calls and blocked my phone number I could not help but wonder if it was due to the influence of her past employer, LAPD playing a major role in what is happening around me since inception, who believes to this day, that although they cannot influence me, that they have a handle on any and everyone around me.

People are consistently told first that the target is seriously mentally ill and not to be believed. Police departments, federal, state and local are now in possession of various types of mind invasive technologies. In my case, every step of the way towards truth, under their monitoring is backed by intense efforts, that others will not discover it.

After telling Richard what I do today, books, etc., and also about my Amazon "Creative Treatment of Actuality" docudrama, "Deceived: Mind Control – Schizophrenic High-Tech Fraud" on Amazon, as I pulled up to drop him off after lunch, right before he exited the car, Richard looked at me with a painful, bewildered look in his eyes and said, "I just don't know what to do about the people talking to me inside my head." "They don't sound like aliens but sound like real people in a building somewhere and they continue telling me what to." Admittedly, this was the last thing I expected to hear from him.

Now shocked, at this revelation, with thousands today reporting Synthetic Telepathy, V2K, or the DOD "Voice of God" harassment and used today at the federal, state and local police level, joined by massive military operations as well, my parting words and hope, was to find some type of logical solace for him. I told him what I know is being reported all over the world by the unleased massive system, which includes devices, documented in the Michigan State laws, and Congressman Dennis Kucinich's Space Preservation Act of 2001 and the awareness that this specific type of nonconsensual experimentation has been going on for years. I did not want to burden him with any conclusion, right or wrong, that there has been a goal by evil in high places, to change everything God created and especially God's highest creation man and woman, by an outside source. I did not want to hurt him or leave any type of destructive conflict to his now chosen lifestyle which he believed freed him. Nor did I want to imply that his decision was against the unconditional love of an omnipotent, omnipresence, omniscient all seeing, all knowing God. His father had already disowned him. This was not my call. My call is to be kind to him, no matter what, as I had done innocently as a child.

His decision was final about the lifestyle change. I felt, why try and tell him that it may have been programming all along to devastate him to the cruelty of a heinous spiritual Coup d' Tat. This program had undoubtedly succeeded in its goal due to his admittance and more importantly his lack of awareness. I told him this is a program and apparently, many, many people have been in it for many, many years.

Without a doubt, now ultimately officially in what appears to be a hope of a final take-down effort unleashed as an adult around me, and a rightful good fight by publication, the realization that I was cleverly set-up for the final stage of destruction of my life was key, as I watched others being taken through various stages of protocol, and it, the result of ongoing monitoring since children.

I also know that as these operations follow specific targets, others are influenced around them proven many times in past and present experiences. I concluded that taking target, who the system has registered as non-programmable, and a system failure, to an official bogus investigation, after legalization, which can officially mobilize stalking, to cover-up ongoing human experimentation is a phase reserved for diehards.

During the year that I was around my dad, broken after divorce, in my mid-30s, after devastating attempts to sexually stimulate me even then, and surely knowing it would fail with me, yet the beam persisted. Why? The devastation is proven useful to this behavior modification program. I believe this well thought out program, feels that there is victory if they create deep, traumatic emotional conflict which results in severe depression, which then contributes to self-medication setting the stage for redirected testing into other areas. The name of the game is to influence self-sabotage.

Today the ongoing malicious, vicious, destruction of families plays a vital role in community destruction by destroying a powerful foundation for positive growth, the family unit, which can set the stage for personal advancement based on a powerful support system. Attempting to traumatize me with BS was happening before I went to the VA hospital which became the bogus foundation for legitimizing their targeting. I was obviously being manipulated before going there by this operation and they had hoped influenced long before I knew this program exists and was 100%, they again hoped set up. They, thus far have succeeded only in this book series. There I was divorced, in deep pain, and this operation persisted in trying to sell to me that something was deeply wrong within the trauma of their beamed negative experiences.

However, my mind would never split because I would not do anything against, or my character, and my will becoming a Sun Wind challenge.

All of this was happening around me, at a time with no awareness of what was really going on then, although today highly educated by failed experience, after experience, after failed experience, thank God!

Later, when I realized the possibility of a possibility, I set out to document everything in "You Are Not My Big Brother..." What unfolded around me should have evolved as a beautiful set-up designed to get me ultimately caught up by the lair of the thought police and military Pops as a continuous effort to take me, and others, to the next traumatic level happening in another major mind control programming site, the corporate prison.

As stated in Book II, the woman who went initially to the LAPD with the bizarre story of a sexual encounter between me and dad disgusted me. I am so much better than that. I had not seen her for over fifteen years, and she as stated in that book, had never laid eyes on my dad. It was later discovered that she was merely a puppet in massive targeting program of her and her family as well.

The fact is, to awaken, I took the Blue Pill which, began to unfold a masterful covert plot, reportedly focusing on thousands, of all races, and some believe millions, designed to understand and fine tuning the global control Matrix.

In gratitude of the truth, as this operation now harassed, tortured, tracked and monitored me I was freed yet again of any lingering doubt. It all finally made sense. Thankful for awareness, I set out on this path to expose this evil operation as the giant, massive, monstrous farce it truly is. If I had been in this program and an outside force had been trying to manipulate and guide me for years, this made the accusations of wrong doing a big fat lie and desire for cover-up for good reason.

Although the little people in the chain of command, used as enforcers, could do nothing about the truth, but follow order and use the technology on me, now under "Kill Orders" for sure, at some level in higher scientific circles and intel agency connections, what is happening today is common knowledge, and it known that every word I speak is 100 percent accurate.

The program appears to shuffle those unbreakable too the final testing arena which also includes rampant sexual stimulation and resulting emasculation of men, especially black men who leave the prison system on the Down Low because of the sexual stimulation. The prison system is a haven for subtle influence for homosexuality which is being created and born from the isolation. These widespread efforts have been a part of covert military mind control programs and military PsyWar nationwide.

The Monarch mind-control experiments reportedly took place at the following locations with sexual exploitation of children historically which plays a major role in trauma-based mind control and dissociation.

Cornell, Duke, Princeton, UCLA, University of Rochester, MIT, Georgetown University Hospital, Maimonides Medical Center, St. Elizabeth's Hospital (Washington D.C.), Bell Laboratories, Stanford Research Institute, Westinghouse Friendship Laboratories, General Electric, ARCO and Mankind Research Unlimited.

China Lake Naval Weapons Center, The Presidio, Ft. Dietrick, Ft. Campbell, Ft. Lewis, Ft. Hood, Redstone Arsenal, Offutt AFB, Patrick AFB, McClellan AFB, MacGill AFB, Kirkland AFB, Nellis AFB, Homestead AFB, Grissom AFB, Maxwell AFB and Tinker AFB

Langley Research Center, Los Alamos National Laboratories, Tavistock Institute and areas in or by Mt. Shasta, CA, Lampe, MO and Las Vegas, NV.

(Project Monarch: Nazi Mind Control by Ron Patton)

It has been widely reported that the US military tortures American children as part of mind- control experiments. With open literature evidence of specific programs, it makes you wonder what is really happening with missing children, today's explosion of Human Trafficking, and at many Top-Secret underground military bases reported across the nation.

Information acquired from the Freedom of Information Act (FOIA) regarding Paul A. Bonacci, said that, as a child, he was kidnapped, tortured and subjected to sex abuse and mind control. In 1999, in a court in Omaha, he won $1,000,000 in damages.

Bonacci in his testimony referred to the involvement of top members of the US military and top politicians involved in child abuse. The Washington Times reported that Paul Bonacci had access to the White House living quarters.

Bonacci also testified on videotape (5-14-1990) for the Nebraska State Police investigator, Gary Caradori:

"Bonacci said that while on a trip to Sacramento, he was forced at gun-point to commit homosexual acts on another boy before he watched other men do the same – after which the boy was shot in the head." A child sex rings is repeatedly linked to top Americans to include, again, today's billions dollar Human Trafficking industry. And shooting another in the head right before a person's eyes is a surefire method of trauma-based mind control and mental programming.

Johnny Gosch is reportedly one of those tortured by the US military. John Gosch disappeared without a trace on September 5, 1982. No remains were ever found.

His mother maintains that Johnny Gosch escaped his captors and visited her with an unidentified man in March of 1997. She claimed her son told her that he was the victim of a pedophile organization. However, Johnny Gosch's father stated that he was not sure the encounter ever happened.

However, in 2006, his mother claimed to have found photographs on her doorstep, on her birthday, depicting Gosch in captivity. Some of the photos were proven to be children taken from an "exclusive members only" porno site and identified by family. However, one of the boys in the photo was never identified. Johnny Gosch's mother, Noreen, insists this is her son.

Noreen Gosch has reportedly received death threats, has been ridiculed, and abused by authorities and laughed at by mainstream media. One thing is certain, she has never been afraid or frightened out of saying that her son became part of a massive, unwitting victim, huge child and human trafficking ring taking over 100,000 kids yearly.

[Excerpt] "Sinister Mind Control Programming"

Source: YouTube

Bryce Taylor, a native California reported that her Multiple Personality (MPD) condition resulted from what she had first thought in 1986 was solely sexual and ritual abuse. But as she began to remember and heal, from more of her hidden past, she realized that ritual abuse was more mind control, trauma-based, inflicted by her pedophiles with close family ties that were used to condition her for participation in Project Monarch and white slavery operations.

Bryce stated that mind-controlled sex slaves are created by starting with a child from birth by inducing trauma, of which for Bryce was done in the form of being sexually abused, and satanically abused and put through a trial of satanic rituals. There were blood rituals where she reports she was forced to witness animals and people being sacrificed. She and others were also taken to the basement for what turned into kiddie porn which was filmed.

Her report, at the end of her testimony, is confirmed as factual by ex-FBI agent, Ted Gunderson, a 27-year veteran of the bureau, and Los Angeles FBI supervisor.

In 1921, in London, the Tavistock Institute of Human Relations was set up to study the 'breaking point' of humans. Intense breaking and dissociation of the mind is the key for mind control and submission.

(Project Monarch: Nazi Mind Control by Ron Patton).

In 1932, Kurt Lewin, a German-Jewish psychologist, became the director of the Tavistock Institute. He studied the use of terror to achieve mind control. One of the images Noreen Gosch reported receiving depicted her son bound and gagged. In Germany, similar research was being carried out by the Germans and many of which, by Operation Paperclip were relocated to the US after World War II.

There are many links between the fascists in Germany and the fascists in Britain to the Nazi programs of the United States and Operation Paperclip and the quest for world dominion and by any monstrous means necessary to starting with breaking down and destroy evolving minds.

The Order of the Golden Dawn, a masonic-style secret society had Aleister Crowley as a member, included German Nazis and British aristocrats.

(Project Monarch: Nazi Mind Control by Ron Patton)

United States military, Col. Michael Aquino, is one of those frequently mentioned in cases relating to child abuse and mind-control.

(Michael Aquino, child sex abuse and the United States.)

Lt. Col. Michael Aquino, Aquino was connected with the Presidio Army Base day care scandal, in which he was accused of child molestation along with some of the key figures holding powerful positions.

In the Judeo-Christian America, the argument goes like this:

The Bible says that homosexuality is a sin (Leviticus 20:13). If sexual orientation has a strong biological component, then gays and lesbians can hardly be held morally culpable. But if it's a choice, then they can be rehabilitated (through "conversion therapy") and forgiven ("love the sinner, hate the sin" goes the popular trope) is the frame of thought. However, what if it is a creation using men, women and children as science projects?

Evangelist Jimmy Swaggart articulated the logic this way:

"While it is true that the seed of original sin carries with it every type of deviation, aberration, perversion, and wrongdoing, the homosexual cannot claim to have been born that way anymore than the drunkard, gambler, killer, etc."

Jimmy Swaggart, with his integrity questionable, would later apologize, after saying he would kill any gay man who looked at him romantically. It was a sexual scandal with prostitutes in the late 1980s and early 1990s that would also lead to the Assemblies of God requesting that Swaggart temporarily step down.

The American Psychological Association defines sexual orientation as "an enduring pattern of emotional, romantic and/or sexual attractions to men, women or both sexes." However, all of the definitions, analysis, and attempt for clarity, and search for understanding fall short if actually the creation is the culprit of a heinous scientific outside source.

Henri Bergson was influential in the tradition of continental philosophy during the first half of the 20th until World War II.

He wrote:

"Fortunately, some are born with spiritual immune systems that sooner or later give rejection to the illusory worldview grafted upon them from birth through social conditioning. They begin sensing that something is amiss and start looking for answers. Inner knowledge and anomalous outer experiences show them a side of reality others are oblivious to, and so begins their journey of awakening. Each step of the journey is made by following the heart instead of following the crowd and by choosing knowledge over the veils of ignorance."

For this, I am most grateful!

CHAPTER FOUR

Id, The Place Where Ego and Libido Meet

According to Sigmund Freud's psychoanalytic theory of personality, the id is the personality component made up of unconscious psychic energy that works to satisfy basic urges, needs and desires. The id operates based on the pleasure principle, which demands immediate gratification of needs. The three major components of personality postulated by Freud are id, ego, and superego.

This foundation of focus on human sexuality indicates an understanding of the power of, by greater understanding of the power of sexual energy in and of itself. Sex is the driving force of nature urging the pollination of plants to the biological urge to reproduce in animals and humans. It is not surprising that that most of our energy arises from libido or id., as a connection of ego, super ego and libido.

For centuries, man has tried to harness and channel this energy into more fulfilling areas of higher states of consciousness or in the case of the "so-called" appointed, delusional Controllers, into lowered states of consciousness for mind control. Our culture today can be defined as a factual sex culture. The result is people worldwide are being encouraged to release their sexual energy which can take the form of various types of aggression based on the urge desire for satisfaction, too include orgasmic aggression or ongoing war programming. Key is to keep humanity operating at a certain frequency.

While randomly speaking with a person on this topic, the individual recounted an incident that at the gym. While working out, he glanced at the television and an old soap operator "Day of Our Lives" was on television. Catching his attention were two naked men in bed and holding each other in a provocative manner as they lay cloaked by the sheet. There then was a pounding on the door and entrance of another man whom appeared to be jealous that the two men were together. The problem here is in the promotion on television.

The fact is constant release of sexual energy is very damaging. In reality, greatness has been accomplished through the ability to curtail and then transmute it into superior spiritual energy. However, bombarded with what should be personal privacy, but it publicized, strategically, this is the last thing those operating in the control Matrix want.

The real power, of the energy surround sexual acts, is in the organic frequency emitted, and more importantly, the power of the frequencies electromagnetic sexual energy to bring forth and materialize life.

When this pristine power is used to achieve beauty, and then tampered with to create an imbalance negatively, it becomes a very effective weapon used against us. Other examples today are found in the breakdown of society, and a massive desire for release of sexual energy which is at the foundation of mass pedophilia and Human Trafficking. This is where implanted thoughts and desires for sexual gratification at any and all cost, steals the beauty. The desire for sexual energy conjuring, can also be derived from pornography, rape, sodomy, which effectively keep humanity operating in an unevolved primal state motivated by primal instincts and trapped in the urges of the animal kingdom. The influence is pervasive and is also why specific types of mainstream music have an overall sexual theme. If sexual gratification is the main thing on the mind there is little room for anything else.

It is important to "Know Thyself" as many have that have awakened, have not been broken, so that you are not prey.

In spite of it all through the maze of life, designed to lead us to our truest self, one thing is certain, I believe that our Creator loves all creations and above all else, the good intent and efforts of the heart reveals the connection with God and good intent for humanity.

A Published excerpt in "The Yale Review" in April of 1926 regarding tin foil / aluminum stated:

"Well, we had discovered that metal was relatively impervious to the telepathic effect and had prepared for ourselves a sort of tin pulpit, behind which we could stand while conducting experiments. This, combined with caps of metal foil, enormously reduced the effects on ourselves."

The oldest reference to tin foil hats, in fact also lies in opposition to the role of secret societies and Freemasonry.

In 1717 there was widespread public concern regarding these group's specific efforts and machinations, as being the ongoing hope for controlling the minds of society.

In fact, those opposed to Freemasons were the first to construct hats made from aluminum foil who then began a fraternity called the "Mad Hatters."

The search for protective deflective materials against electromagnetic energy psychophysical attacks continues today. Thousands of highly credible people in this search are being discredited as well.

However, the "Mad Hatters" in the 1700's believed if you were wearing a tin foil hat, you were impervious to the mind control of Freemasons.

As I continued my research, interesting, I learned that the term "Mad as a Hatter" is linked to the madcap milliner in Lewis Carroll's classic children's book, Alice in Wonderland and Alice's journey down the Rabbit Hole or death or unbirthing, of the true self is an Anachronism of the Ages.

The Mad Hatter is seen when Alice wonders off in the forest and the Cheshire Cat tells her to visit the Mad Hatter and March Hare for directions back home. Alice visits as the two are in the middle of a very odd tea party with the Dormouse.

When Alice finds them, they are singing "The Unbirthday Song" (death to self) and are interrupted when Alice applauds and compliments which creates a bond.

"Unbirthday," described in the Urban Dictionary, is a symbolic meaning celebrating a future time of death. If viewed in the dynamic of mind control, it can be related to mental trauma and death of the true self. Mind control programming is simply creating a new world for the person they are engaging and an alternate reality within and fractured mind.

Monarch programming techniques focus on dissociation, and is documented to be part of trauma-based mind control projects, specifically spearheaded, as yet another study of many, by the Association of Psychiatry, and documented by Dr. Colin Ross, known as Project Bluebird. In Monarch programming, the programmer takes the subject in their broken down and susceptible state and begins to program in a new world or reality with words, used as triggers, music or sound. The term "unbirthing" used in the tale of Alice in Wonderland, is a key word.

Walt Disney was one of the secret society families. The fact that Walt Disney chose the name of "Alice" as his first work tells us about Disney and some would argue, cleverly concealed symbolism of trauma-based mind control as a creative treatment of actuality. Symbols and symbolism are the keys to the structure of the steps of Freemasonry.

Both Alice in Wonderland and the Wizard of Oz are about little girls. Both in the dramatization spend much of their time avoiding danger. In the case of Alice, she gets kidnapped by the cartoon villains, or Dorothy is targeted by a wicked neighbor who then becomes the Wicked Witch of the East.

Alice is threatened with trauma, or an unbirthday signifying the death theme, such as being tied to a log in a sawmill.

Themes creating dissociation are clearly present in both Alice in Wonderland and the Wizard of Oz and also many modern tales.

Alice in Wonderland is about a little girl's dissociation, and specifically from sexual abuse, and little known with a pedophile connection. The fact is, speculation about Lewis Carroll, the writer, later arose which questioned whether he was actually factually a pedophile. Discovery of a cryptic letter he wrote raised questions about the author's relationship with the 'real-life Alice Liddell' known coincidentally as 'little' Alice in Wonderland, as well as the possibility of Lewis Carroll, who really was named Reverend Charles Lutwig Dodgson's personal involvement with her.

The real Alice Liddell's father was the Dean of Christ Church and Reverend Dodgson (Lewis Carroll) was a close friend of the family. This was until some type of secretive separation and halting of the relationship in 1863 and Dean's daughter, Alice then 11 years old. Three entries from June of 1863 were cut out of Dodgson's diary holding details of the split. And after the missing pages disappeared Reverend Dodgson did not socialize with the Liddell's anymore.

A letter, by Reverend Dodgson was written in a room at the Christ Church College in February 1877, when he was 43 and Alice was 25. In the letter Dodgson writes of a preference for little girls rather than boys to a protégé.

The question is did Dodgson act on his desire for young girls while living with the Liddel family? Did both Dodgson take himself and little Alice Liddel down the rabbit hole of his desires, urges and fantasies?

Later he himself, after revelation of what he had written became in his diary become public, he stated that the note was taken out of context.

Dissociation is a basic survival mechanism resulting from extreme trauma of an abused child resulting in the mind "splitting" or symbolically shown in these two, appearing harmless, Disney productions. To those who understand the goal of trauma-based mind control, going from a conscious reality reflected by tumbling down the rabbit hole, or Dorothy awakening in the Land of Oz indicates two extremely different realities.

Alice is led by the White Rabbit representing her Handler who is also, if a possible connection is understood, connects the writer of the story and the possibility of a real-life connection about whom the story is actually about.

The question is, was "Alice Liddell" a victim of drug influence and resulting sexual abuse?

Symbolically, Alice's, programmer, Carroll takes her into Wonderland resulting from the mental split and devastating change of her reality. Going through the looking glass or mirror becomes the method of dissociation and separation and the consequent of emerging two separate selves.

It appears to be a subtle revelation in these two productions of the awareness of the human brain's capability for redirections, hypnotic mind control, and blocking of memory. Alice remembers only reality in a dream. And, within the theme of Alice in Wonderland is a thinly veiled, association with drugs and the role drug's play in the journey down the rabbit hole assisting in dissociation.

It's a tempting theory. After all, the story has other drug references such as Alice eating 'magic' mushrooms and meeting hookah smoking caterpillars. Also, during this period opium and Laudanum, which is tincture of opium by weight was in full use. Laudanum contain almost of the opium alkaloids, including morphine and codeine. Some reports hinted that Carroll may have been involved with drugs, although unsubstantiated.

There was also a dark side to Carroll. He suffered from a strange disorder that caused him to have hallucination's which made him feel bigger or smaller than he was. And this theme feature is prominent in the story that it subsequently became known as Alice in Wonderland Syndrome.

The perception a person can have due to micropsia, a potential symptom of dysmetropsia. From Lewis Carroll's 1865 novel *Alice's Adventures in Wonderland.*

Alice in Wonderland Syndrome:

Alice in Wonderland syndrome (AIWS) is a disorienting neuropsychological condition that affects perception. People experience size distortion such as Micropsia, macropsia, pelopsia, or teleopsia. Size distortion may occur of other sensory modalities.

It is often associated with migraines, and the use of psycho active drugs. AIWS can be caused by abnormal amounts of electrical activity causing abnormal blood flow in the parts of the brain that process visual perception and texture.

In the excerpt below The Rape of the Mind: The Psychology of Thought Control, Menticide and Brainwashing by Joost A. M. Meerloo, describes menticide as a factual form of brainwashing.

Meerloo writes.

"Menticide is an old crime against the human mind and spirit but systematized anew. It is an organized system of psychological intervention and judicial perversion through which a powerful dictator can imprint his own opportunistic thoughts upon the minds of those he plans to use and destroy. The terrorized victims finally find themselves compelled to express complete conformity to the tyrant's wishes."

He further states:

"The totalitarian potentate, in order to break down the minds of men, first needs widespread mental chaos and verbal confusion, because both paralyze his opposition and cause the morale of the enemy to deteriorate unless his adversaries are aware of the dictator's real aim. From then on he can start to build up his system of conformity."

There has always been hierarchy diffusion for ongoing social and mass population control. Structures of religion, education, set the stage backed by highly effective intentionally created manipulative propaganda, and powerful media influence, to include drugs as well as the advanced, bioelectric advanced technology being revealed today anew.

Through rearranging our perceptions, the hope is to hold humanity in a holographic reality and stop and/or limit the organic ascension of the soul's natural progression which naturally demands the advancement of human consciousness and Spiritual Awakening.

Our hologram is composed of grids created by a source of consciousness brought into awareness by electromagnetic energy at the physical and material level of this plane.

Hierarchical diffusion is an idea or innovation that spreads by moving from larger to smaller places, from the top downward so to speak, without concern to the distance between places, and is often influenced by social elites.

Contagious diffusion refers to the widespread diffusion of a characteristic throughout the population.

Since birth our mass consciousness is formed and created. Everyone around us and everything has been indoctrinated by propaganda, of both covert and overt influence. Most following the systemic structure, blindly for guidance never wake.

The fact is we are born into a system that conditions us from birth and continues throughout our lives by intent and the Handlers use any method they can with an understanding of the bioelectric mind and body as well.

In the opposing duality, of self and intervention, and unification, a clear invitation lies ahead for us. It is to see who we are and how our past definitions of ourselves have been radically altered many times negatively. The end result of the awakening brings wholeness.

There is both rhyme and reason for ongoing human experimentation. The cruelty or torture is beyond belief and so deeply disturbing that many have concluded and attributed it to a level of hideous, hidden acts founded by those operating within the realm of satanic obsession and possession.

Looking back thousands of years reveals that the Mystery Religions of ancient Egypt, Greece, India and Babylon helped to lay the foundation for occultism practices today by secret societies. Within the Egyptian Book of the Dead, can also, be found a compilation of systematic rituals explicitly describing various methods of torture and intimidation which are specifically designed to create trauma. Trauma clearly reprograms, the mind can then be transformed and recreated then used for a specific purpose or idle with no real advancement dormant.

The book also details the use of potions (drugs) and the casting of spells (hypnotism) used as techniques which specifically result in total enslavement of individuals and on a larger scale a population, for example by continuous unleashing of drugs, again, both legally and illegal globally. These capabilities have been the main ingredients for a specific part of occultism known as Satanism, throughout the ages.

The history of Satan is described in the Bible in Isaiah 14:12-15 and Ezekiel 28:12-19.

These two biblical passages also reference the king of Babylon, the King of Tyre, and the spiritual power behind the Kings. The Phoenician king of Tyre appears in the Bible as an ally of both Kings David and Solomon. Solomon is derived from the word sol, om and on. The meaning of the word sol is the second sun, light, soul and peace.

The Winding Staircase in Freemasonry is connected to "Solomon's temple, or the perfected temple of the human body, the perfected temple of the universe and the perfected building Solomon's Temple.

Manly P. Hall in his great book, the Secret Teachings of all Ages calls Solomon, The Spirit of Universal Illumination-mental, spiritual, moral, and physical. The name Solomon he states, symbolizes light, glory, truth collectively and respectively. It signifies its invisible but all powerful spiritual and intellectual Effulgence."

The Source of the Effulgence, and the Effulgence, and the Recipient of the Effulgence; the so-called Illuminator, and the so-called Illuminati, are therefore the Illuminated.

[Excerpts]

"...temple of the soul finally forms the perfect shrine for the living Ark."-The Initiates of the Flame Manly Pl Hall, Pg. 59

"This temple (Dionysian's) was human society perfected; and each enlightened and perfected human being was a true stone for its building."– The Secret Destiny of America by Manly P. Hall

"...true temple of Solomon is the universe, the Solarman's temple, which is slowly being rebuilt in man as the temple of the Soul of Man."-The Initiates of the Flame Manly P. Hall, Pg. 81.

Most are familiar with the biblical account of what caused Satan to be cast from Heaven? Satan fell because of pride that originated from his id desire of ego, superego to be God instead of a servant which is typical of the intensity of ego driven desires harnessed today.

Today Satan is often caricatured as a red-horned, trident-raising cartoon villain or Baphomet.

But how much influence as Satanic influence had on evolving a decisive "New World Order," depopulation, and world dominion by Globalist? As stated, previously the task of mass, social and population control requires formidable awareness of mind control and advanced understanding of the bioelectric effect of the Electromagnetic Spectrum.

Satan symbolic image and existence, when depicted as caricature is not based on fantasy. Many structured religious beliefs document that Satan acts as leader of the fallen angels by energy. These demons are said to exist in the invisible spirit realm yet affecting our physical world or as biblically reveals another dimension of rebels against God.

Satan also deceptively, masterfully masquerades as an "Angel of Light" and as the illuminated one coincidentally connecting the name Illuminati to this negative deceptive energy source and motivation. Interestingly in the bible Jesus and Satan are both referred to as the Morning Star.

The verse in question about Satan is in Isaiah.

How you are fallen from heaven, O morning star, son of dawn! How you are cut down to the ground, you who laid the nations low! **(Isa. 14:12)**

The verse about Jesus is in Revelation.

I, Jesus, have sent my angel to testify to you about these things for the churches. I am the root and the descendant of David, the bright morning star. **(Rev. 22:16)**

It's clear from the context that these two passages aren't talking about the same person and this is if Satan is actually a person. God certainly is not.

In the context of Isaiah 14, Satan is referred to as the King of Babylon. Some don't agree and say this Isaiah passage isn't about Satan but is indeed about the King of Babylon.

Even if Isaiah is calling Satan "morning star" he is "a" morning star, not "the" morning star!

During the 13th Century, the Roman Catholic Church increased and solidified its dominion throughout Europe with the infamous Inquisition. Satanism survived this period of persecution, deeply entrenching itself under the veil of various esoteric groups and of which continue to thrive secretly today.

Today, the ongoing war status, around the world, unfolds, as a connection to secret esoteric groups, fueled by what some conclude as thousands upon thousands of deaths as actual satanic blood sacrifice for the Luciferian agenda.

The bible further states that when the "Sons of God," and Angels, presented themselves before God, Satan was there and a conversation ensued about Job's goodness. Satan challenged God Almighty by stating that Job would denounce God if afflicted.

God gives permission to Satan to afflict Job. We know that Job doesn't denounce God. So, the question is why would God allow Satan to do this?

Is the reason so that God may be vindicated at God's word and so that we might understand that trials and tribulations will come to those who are godly? I believe that many on this path of our lives, have been prepared and strengthen for this battle today, up against pure evil working through others. The battle is bringing it to light.

An old man we fondly called Grandpa told me during a difficult time in my life. "You are being tested." It is amazing how words can also free you. In fact, it was many tests that laid the foundation by strengthening me to fight today in battle for my life.

One thing is certain, the human brain plays a vital role in consciousness and is the antenna for consciousness and programmable. This is something which must be fully grasped, in this awareness, for protection, hold onto and never lose sight of.

The possibility of advancing evil, by many strategic methods, producing of an unbalanced brain consciousness, is designed to keep people in both a mental and spiritual schism. This is also depicted symbolically by the black and white checker board imagery of Freemasonry.

Alice in Wonderland Checker Field

Freemasonry Symbolism

Land of Hypnosis

Even the colored image of this scene has the black and white checker board background still in black and white.

Black and white is a symbol of duality, used in ritualist ceremonies, and also represents the base consciousness or in the case of mind control, transformation into something new. Mind control works by continuously, dividing these two parts of the brain or the qualities that are active in these two parts of the brain by continuously trying to keep them separate.

In the 1960s, researchers at the University of California began an experiment to study changes in blood pressure and blood flow. The researchers used 113 newborns ranging in age from one hour to three days old as test subjects. In one experiment, a catheter was inserted through the umbilical arteries and into the aorta. The newborn's feet were then immersed in ice water for the purpose of testing aortic pressure. In another experiment, up to 50 newborns were individually strapped onto a circumcision board, and then tilted so that their blood rushed to their head and their blood pressure could be monitored. These studies reveal the callous perception of young life.

In 1965, Canadian David Peter Reimer was born biologically male. But at seven months old, his penis was accidentally destroyed during an unconventional circumcision by cauterization. John Money, a psychologist and proponent of the idea that gender is learned, convinced the Reimers that their son would be more likely to achieve a successful, functional sexual maturation as a girl. Though Money continued to report only success over the years, David's own account insisted that he had never identified as female. He spent his childhood teased, ostracized, and seriously depressed. He was yet another science project for the criminally insane. At age 38, David committed suicide by shooting himself in the head.

According to a small note by the Washington Post, the CIA also financed a study to evaluate psychological consequences of circumcision, trauma, later to be identified as MKULTRA Subproject 74.41. The Post disclosed on October 2nd, 1977, that the Central Intelligence Agency funded experiments in the early 1960s on circumcised children to determine if the operation left any emotional after effects.

The aim was to determine if circumcision at a significant stage of a child's development produced anxieties such as fear of castration or the inability to resolve a castration complex and its link with later emotional disorders, notably homosexuality, said a CIA research paper on the experiments.

In 1980 Edward Wallerstein published a book about circumcision that included the following passage:

> "Probably the most outrageous circumcision study of the century was reported in the New York Times in 1977. The Central Intelligence Agency reported that in 1961 it had arranged to have 15 boys, aged 5 to 7, circumcised. The boys were from low-income."

Unethical human experimentation in the United States describes numerous experiments and race is not the motivating factor, no one is exempt.

Such tests have occurred throughout American history, but particularly in the 20th century. The experiments, documented by open literature evidence, include many tests were performed on children, the sick, and mentally disabled individuals, often under the guise of "medical treatment." In many of the studies, a large portion of the subjects were poor from rural areas, racial minorities, or prisoners.

Whitey Bulger, a former organized crime boss, wrote of his experience as a prison inmate test subject in MKULTRA, confirming years of prison experiments.

He wrote:

> "Eight convicts in a panic and paranoid state," Bulger said of the 1957 tests at the Atlanta penitentiary where he was serving time, reported total loss of appetite and hallucinating. The room would change shape. Hours of paranoia and feeling violent" which we know today can be achieved by frequency manipulation.
>
> We experienced horrible periods of living nightmares, he wrote and even blood coming out of the walls. Guys turning to skeletons in front of me. I saw a camera change into the head of a dog. I felt like I was going insane."

These human experimentation programs, of which were many, and ongoing, laid the foundation for what is happening today. Many of these programs as well have been the result of military spearheading massive social engineering for globalist making you question, who really owns the military?

Reportedly, the Ronald Reagan UCLA Medical Center was home to top mind control programmer and psych ward head, Louis Jolyon "Jolly" West. It is believed that his work continued well into the late 80s.

Between 1974 and 1989, Jolly West received at least $5 million in grants from the federal government, channeled through the National Institute of Mental Health (NIMH), a major funding conduit for CIA programs. The money continued to flow into the Neuropsychiatric Institute of which West headed growing to 14 million.

West did his psychiatry residency at Cornell University, an MKULTRA institution and site of the Human Ecology Fund. My point is that nonconsensual has a deep-seated heinous reality in California specifically.

Jolly West later became a subcontractor for MKULTRA, Subproject 43, by a $20,800 grant by the CIA while he was Chairman of the Department of Psychiatry at the University of Oklahoma.

The proposal submitted by West, a specific focus of his research was titled the "Psychophysiological Studies of Hypnosis and Suggestibility" with an accompanying document titled "Studies of Dissociative States."

So-called Conspiracy Theorist, report that UCLA has remained the main programming site for celebrities, Kanye West, Britney Spears, Amanda Bynes, Michael and Paris Jackson, and Lindsay Lohan to name a few. Whether fact or fiction, efforts focused on programming cannot be denied confirmed by years of open literature evidence and so diabolical it is difficult to digest.

Controversy arose when, in the summer of 2018, his wife reported that Kanye was no longer the man she married. And some believed it hinted that after his so-called reprogramming, after his original programming failed, said to be vital for all puppets of the entertainment industry, which includes sports figures that he became a deficit as a money maker resource for the music industry elite. After he malfunctioned, he began to spill the beans, was out of control and had to silenced and reprogrammed. As for me, I simply say fact or fiction, you decide.

Oddly when he returned to the scene, he then shocked everyone by outrageously saying that 400 years of slavery was a choice.

This is so absurd, insane, and illogical that some would argue it must be new programming. How else would he be allowed the promotion of this pure nonsense, reeks of a programmed mentality of such nonsense by his Handlers, while making a complete fool of himself above all other comments? You must admit that the Illuminati controlled entertainment industry and controlled media would want a programmed puppet with media influence to report, in the hope to continue to entrain minds and perceptions with disinformation, a denial of historical facts.

The use of the media for strategic propaganda also includes relentless, mindless distractions which effectively steer our attention away from the global horror being inflicted right before our eyes.

After his last psychological breakdown, and him dying his hair blond, it too could be interpreted as a sign that he had been effectively reprogrammed, and once again, so to speak, white-washed by his Handlers.

Many agencies have been mentioned, but without a doubt, leading the way today in the global paradigm is the military, military technology and specifically the United States Air Force.

While the Army, are foot soldiers, Navy sea warriors, and the Marine Corps also known as the Naval Infantry are typically an infantry force that specializes in the support of Naval and Army operations, the USAF and its activities have been long cloaked.

"Stranger Things" the Netflix, series is a fairly tense, very 80s, sci-fi thriller detailing spine- chilling stories at Camp Hero air force base in Montauk.

[Wikipedia]

The Montauk Project is an alleged series of secret United States government projects conducted at Camp Hero or Montauk Air

Force Station on Montauk, Long Island, for the purpose of developing psychological warfare techniques and exotic research as an outgrowth of the Philadelphia Experiment.

As the program continues its focus me USAF personnel are without a doubt, the main ones I witnessed setting up operations in specific locations, each time I moved in the past, however, as stated earlier, I did also see FBI entering locations to include the Los Angeles Police Department as well and still do. And, the operation around me, in California falls, under the guidance of the Joint Resource Intelligence Center (JRIC) the FBI counter-terrorism division with some targets strategically labeled as "Domestic Terrorist" approving the advanced technology testing

The below patent is mentioned in two of the books in this series.

This is because; I believe it vital to understand it capabilities and full use today The Abstract for

US Patent, #5, 159.703 reads:

A silent communication system in which non-aural carriers, in the very low or very high frequency range or in the adjacent ultrasonic frequency spectrum, are amplitude or frequency modulated with the desired intelligence and propagated acoustically or vibrationally, for the inducement into the brain, typically through loudspeakers, earphones, or Piezoelectric transducers. The modulated carriers may be transmitted directly in real-time or may be conveniently recorded and stored on mechanical, magnetic or optical media for delayed or repeated transmission to the listener.

Here's the clincher,

According to literature by Silent Sound, Inc., it is now possible using supercomputers to analyze human emotional EEG patterns and replicate them, then stored in "emotion signature clusters" on another computer, and at will, silently induce and change the emotional state of a human being.

When speaking about drones, equipped with psychotronic weapons, the Sound Spread Spectrum System (SSSS) is among the arsenal. It is the force used while tracking a target everywhere and having the capability to wreak havoc by frequency influence on people around targets of whom these operations hope to use covertly against the target through beamed subliminal influence.

It would be pure science fiction if the police were not in fact using 'Thought Crime' predictive policing technology today. Today it is science, nonfiction.

In George Orwell's 1984, the "Though Police" or "Thinkpol" has grown as a high-tech reality for many major police departments, nationwide having advanced secret police divisions which today, are familiar to LAPD and for year, the FBI.

Stores I frequent in my community, the gym, have become havens for beamed subliminal influence of those around me and using people by Remote Neural Monitoring, without the operatives leaving the operation center building. The goal is to get a negative reaction thus justifying the mental illness tag after the target reacts, in my case strategically, after added to my medical records, a few months before this book's publication then try to set me up.

When I went to the VA hospital and reported honestly that the FBI, LAPD and military in a joint targeting effort are after me, in a focused targeting program, apparently this as reality is so outrageous, that it must not be true, twelve years later and many credible publications.

Perhaps one of the best examples of the SSSS patent influence was an incident that happened while working at Starbucks working on this manuscript. I had taken several images of the drone following me seen clearly at night tracking me home then positioning itself stationary over my house each night and finally caught an image of one during the day which day tracked me to Starbucks and positioned above.

I had gotten there early one morning, working all day, and around dinner time, with still more to do before calling it a night, I decided to purchase food so I would not have to leave and break the internet connection vital for what I was downloading, which was my Amazon DVD.

When I approached the cashier to place my order area, the store manager was putting together a sample tray of appetizers to hand out to customers. I glanced at her and said "Hi."

She then said, "I was going to offer you one of my ham cheese rolls, but you're Vegetarian." Surprised, she knew this about me, I asked her, how did she know? She in turn was also surprised. Dumbfounded she then stated that she heard it around from the area, unbelievably, of which I was sitting a good 40 feet away from where she stood. I had been sitting there working with headphones on working.

This personal information about me is unpublicized with my having no need to tell folks how I eat. It was never discussed with any stranger, so there was absolutely no reason for her to know this. She shook her head confused searching for another logical answer of how she knew and there was none. She looked totally confused.

I already knew and started to tell her that it was likely a beamed subliminal thought by the drone overhead, and the character's using the technology, hypnotized on false importance because of the capability, but because it was happening everywhere I went, I felt it pointless. Also, I realized that awaking some people could result in them also becoming a targeting focus and on this day, I just did not want to go there which would require my taking the time to break this program down. I decided to let sleeping dogs lie for now with her.

It got better. When I returned to my seat, a stranger came in and sat next to me. For some reason, he could not stop staring at me then quickly looking away. Finally, when it happened so many times, I asked him why? He shook his head, saying he did not know but stated that he heard someone else talking around me. Without a doubt this was the drone operator, again by the beamed communication system. They were asking him was I good looking. And from experience, may have been even trying to incite fear in me by using him by their constant threat of having someone rape me.

Everywhere I went, if I were not aware of the program and the technology I could have misinterpreted reactions towards me as insulting or threatening.

And with this happening rampantly, I also had to consider that it was also likely the motivating source and beamed influence that brought about the mental illness tag addition to my medical records as I mentioned by a Moslem Resident at the VA hospital.

I highlight Moslem, because one thing is certain and documented, in the "Surveillance State," and which has included ACLU lawsuits brought on behalf of the Moslem American community, there is an ongoing focus on them and continued monitoring from nationwide counter-terrorism divisions searching for any possible hidden terrorism and these division are fully equipped technologically. This is where technologies like Hewlett Packard's Pre-Crime and the Department of Homeland Security's Malintent come into play with little publication of the capability to decipher thought.

In reality, by July of 2018, the operation embarked on what appeared to be a final effort of silencing. Noted earlier, the focus on me included beam cooking of specific tissue, organs and joints.

The operation hated the fact that wherever I went I was warmly greeted, welcomed and treated well based on how I have always carried myself. The key to this program from a stalking perspective is to first send out federal agents, to set the stage providing negative information about the target being, again, severely mentally ill, a prostitute, thief, criminal, pedophile, child abuser, etc., then after the thought is implanted by the physical visit, the operation center takes over and everyone is monitored from that influence, so they don't forget, while the target is tracked and interacts with people. The reported allegations replayed over and over to the person, after the initial contacted.

I could not go into the water store which I frequented for a few years, and spent a lot of money there, without the owner acting like I was stealing something.

Every time I touched, the perfume she sold, although I had bought two bottles from her, she walked up behind me and checked to make sure it was still there. I knew, again, from repeated experiences, there was only one place this influence was coming. As a result, I almost stopped going to her for business because of the obvious influence and repeated insult. However, it was happening so many times, everywhere

I went that I realized it was not her fault but those at the helm of the technology. Plus, the alkaline water purchased from her store has contributed to a large degree to my health maintenance as the irradiation of my body sought to destroy the balance and throw my body off balance which could strategically result in effectively beamed, high-tech Cancer. Alkaline is key as is drinking lots of water to replenish from the microwave beamed energy weapon depletion.

I knew the stage had been set and it confirmed one day when I pulled up and actually saw an operative, who arrived minutes before me, who arrogantly thought I would not recognize him, talking with her then leave without a purchase, and attempting to not have eye contact with me as I entered the store.

These operations know where you are headed as they watch you dress by use of mind reading technology deciphering your every move.

Jealousy is also another tactic these operations use around targets.

Jealousy is an indicator of deep pain which some carry after having separated from the essence of the true self, exploited by these operations as they watch others shine in the fullness of theirs. It is also extremely useful due to its maniacal motivation.

When I first became a target my beautiful older daughter of three was the main person around me. When she started talking about becoming stripper and prior too, reporting she was hearing men harassing her from the ceiling inside her apartment, I later learned that it was likely the Los

Angeles Police Department, dirty cop division, also detailed in a different book entitled LAPD's Secret Police, on Amazon, by a once insider, erroneously, if you ask me called Elite.

After all, as documented, and stated several times, the fact is that the woman, who ignited their unleashing on me had gone to them first then transferred to the FBI with the military, again simply part of the new targeting model.

When hints of a possible attempt to use this daughter arose, it would be the stripper comment that drew my attention and my wondering where the idea originated.

Later when the hope to use her was beginning to formulate, reported to me by someone who contacted me, it was clear their hope was to use her for Human Trafficking or as their sex toy. You can bet I put a stop to this before it was able to materialize. Combined with an understanding of the beamed sexual stimulation capability around me, and numerous targeted women reports she would have been toast to them. An attempt was made to use this girl in more ways than one. She likely had been set up to be blackmailed, this way for use against me in the set-up, by this wretched group of men.

For over ten years, things got so bad between me and my baby, to the point that I had to break ties with my oldest daughter, who now thank God is an adult with a family of her own. The effort persisted by blackmailing her after setting her up, and her strategically turn against her. They just would not let her be. Had it not been for her sisters, of three girls, not cosigning her programmed agenda she would have long been used as a weapon against me.

It was she who had tried to factually have me committed twice, once in 2010 and again in 2014 for this operation.

When she reported in a Facebook post that she had been raped, I knew that it was not farfetched that those involved in this program targeting, again seventy percent women, or those living alone, with this girl a looker, they could have been involved. They without a doubt were watching her inside her apartment, showering I noticed when I visited as I heard the optical lens of the see through the wall technology focus as she showered over the shower one day.

In the early stages of my official targeting, and as mentioned extreme sexual stimulation, the hope for sexual gratification, detailed in Book II had literally brought one of them to my door.

These men sit and watch targets, day in and day out and if beautiful women, all the better for them who then can operate like a pack of mentally disturbed wolves. In fact, as I typed this portion of what is happening around me, the drone weapon, as they watched, began yet again intense cooking of my now lungs and my heart muscle and yet again, moved my skull, then breast and hip tissue area. It was supposedly their message that we are going to ice you so you better not report this and stop.

If so, it would not be without my first telling the world of these pathetic Human Monsters, of all levels in this program, where literally, "Birds of a Feather Flock Together" literally.

In a moment of weaknesses for a child that I love I contacted her in early June of 2018. The effort around me became a renewed hope to use her, hoping to bank on the recent medical record insertion, with her again, trying to use one of her sisters for support and all girls in another state.

I prayed for the best for her and God's protection of all three daughters, knowing full well these operations will go after your family, as I walked this determined path and carry on.

They also tried to get her, for discrediting, to portray her as bogusly child abused, which was amusing. This was the first born, and when I say super spoiled, it is an understatement. I had supported her well into her late twenties, any and every possible way I could which became financially draining.

This child, if falling her feet could not touch the ground without my catching her before. Her sisters knew this was not true so when she called, her younger, under the prompting of the LAPD's continuous effort and motivation, and tried to reel her sister, she called me and said she was not going for it and that her sister must be crazy. She is not, just being blackmailed. A songwriter, she even hinted at her encounter with them in one of her popular songs.

Again, if a person is targeted since a child, this truth reveals that all of the cleverly constructed narratives, these operations sell to people to gain public trust and support for this evil technological program and used to justify strategic technological torture and murder, negatively is a fake official operation. Instead it reveals these operations for who and what they factually are, liars, opportunist, bullies, immorally seeking control of anyone and perceived above the law and using and abusing women to feed their hunger for importance, gratification and grossly deranged, distorted egos. This program is engulfed by a powerful evil energy and deep hopeful programming.

In 1992, Dr. Corydon Hammond, a Psychologist from the University of Utah, delivered a lecture entitled "Hypnosis in MPD: Ritual Abuse" at the Fourth Annual Eastern Regional Conference on Abuse and Multiple Personality, in Alexandria, Virginia.

(Project Monarch: Nazi Mind Control by Ron Patton)

Hammond referred to the Nazi connection and military and CIA mind control research. Again, he is one of many who have courageously spoken out risking destruction their lives and careers.

As stated previously, among the key figure of Canada's MKULTRA effort was Ewen Cameron's intimate named Alfred C. Kinsey,

Alfred C. Kinsey, an American sexologist, in 1947 founded the Institute for Sex Research at Indiana University.

He co-authored the highly influential Sexual Behavior in The Human Male, (1948) and Sexual Behavior in the Human Female (1953) becoming a controversial figure in the 1940s and 1950s. His work is said to have been a major influence on sexuality within the United States and globally.

While teaching at Indiana University, historians believe his work in cultural deconstruction would ultimately succeed in decimating American sexual mores, help to fragment the family, and would leave the population far more vulnerable to reproductive, cultural, familial, and mind programming.

Kinsey visited the occultist Aleister Crowley, "the Great Beast," known by the press as "the wickedest man alive" at Thelema Abbey shortly before his death in 1955.

His reported goal was his obsession with obtaining the Great Beast's day-to-day sex diaries for use in obtaining grant monies and to maintain the support of the university. In reality, history documents that Kinsey factually needed the excuse of research to validate his twenty-four-hours- a-day obsession with sex.

Kinsey was bisexual and, as a young man would punish himself for having homoerotic feelings. He and his wife were one of the first open marriages. He had sex with men one of which reportedly was his student, Clyde Martin.

Kinsey's most effective weapon for psychological warfare on the world was his Rockefeller- funded study of American sexuality, the most famous volume being his vaunted study of "normal" sexuality, Sexual Behavior in The Human Male, published in 1948

According to Judith A. Reisman, PhD, the author of Kinsey: Crimes and Consequences, "Kinsey's quantitative research, and his numbers, were a perfect fit for Rockefeller to utilize the mass media to 'shape public attitudes and conduct.'

Attitudes are also subliminally changed through mass communications, which caused a rejection of chastity, self-control and moral public governance, as well as promotion of increased illicit sexual conduct.

'Social management' of this sort was nothing less for Rockefeller than changing America's way of life, by among other things altering what Kinsey would call 'breeding patterns' along an evolutionary or the animalistic view of human sexual conduct."

Kinsey's research included observation of child sexuality, the manual and oral stimulation of children's genitals, and the timing of child orgasms with stopwatches. Part of Kinsey's collection of sex films included films of children in sex acts and adult-with-child sex.

According to Reisman, "The Kinsey Report claims at least '317 pre-adolescents' were sexually experimented upon by 'older adults,' and confirmation of at least 2,035 child experimental subjects were later admitted in 1980 by assistants Gebhard and Pomeroy as reported in "Ethical Issues in Sex Therapy."

Another reason for this is the decoupling of sex from procreation, thus making the institution of sterilization and birth control policies, including abortion, more easily introduced to the masses as a covert Eugenic agenda.

However, the forces whom have sought to limit real freedoms have historically also been the first, to cleverly exploit the destruction of traditional values specifically and focus on destruction of the family unit of which black people especially have been hit. There are 3.6 million black women single by design.

Decoupling is especially true regarding sex and exemplified by the Nazis, for whom, according to author Pearl Buck, "Love was old-fashioned, sex was modern." It was the Nazis who restored the 'right to love' in their propaganda, writes Buck, yet ironically it was also the Nazis who also condemn homosexuality and crucified homosexuals while taking sexual conduct to the level of depravity.

"Hitler and His Drugs: Inside the Nazis' Secret Speed Craze" (Rolling Stone, March 15, 2017)

The article focuses on author Norman Ohler, book entitled 'Blitzed,' which describes a drugged out Nazi party and how methamphetamine, oxycodone and morphine fueled the Third Reich.

Where there are drugs there also is mind and sexual programming and perversion.

[Wikipedia Excerpt]

"Bestiality and Germany"

Deceived Beyond Belief the Awakening

In 1932, America and Europe were still reeling from the depression. This was reflected in Weimar Germany's "Cabaret Berlin" when nude "straight" and "gay" dance hall entertainments and drugs dominated the urban cultural landscape. Weimar Germany's wide-open pansexual revolution preceded and, in fact, laid the groundwork for the National Socialist (Nazi) takeover. This also mirrors the Hippie "Free Love" movement in the United States now under years of Nazi mind control experiments continued in the USA, as stated by recruitment after World War II.

As in the French and Russian revolutions, Germany's political upheaval would also be preceded by a sexual revolution, with thousands of destitute boy and girl prostitutes, fascist and communist youth roaming the violent streets in search of customers and recruits. Alex de Jonge, writes in "The Weimar Chronicle: Prelude to Hitler," that post World War I widespread inflation destroyed the stable, conservative middle class, and predisposed its youth to cynical ruthlessness and disorder.

The resulting "trauma destroyed savings, self-assurance, and a belief in the value of hard work, morality and sheer human decency...." "Traditional middle-class morality disappeared overnight. People of good family cohabited (consider that it was illegal to do that and had illegitimate children..."

Nazis restored the 'right to love' in their propaganda," creating a new decadent and dissolute generation that put Berlin on the cosmopolitan pleasure seeker's map." (p. 295)

"Exploiting this revolutionary upheaval, Hitler had recruited and trained his 'Hitler Youth' since about 1922. Adult males seeking youthful boy consorts traveled to Berlin from all corners of the world to join in on the excitement of the wide-open German free-sex movement," Judith Reisman continues. [p.296] [11]

Let us hope and pray in awareness that bestiality now sweeping across Europe, with sex sheep farms, and the foundation of the Operation Paperclip mind control program transfer to the United

States and influence years ago still prevalent today does not continue it atheism, and moral decay.

It is imperative to grasp, again, that the whole objective is, again, and again, has always been to strip humanity down to something animalistic and primal and to keep you this way.

never let a STUMBLE IN THE ROAD BE THE END OF THE journey

CHAPTER FIVE

Deceived Beyond Belief

The "Malleus Maleficarum" treatise describes ways to manipulate a person into false confession. In reality, the Malleus Maleficarum or "Hammer of Witches," was the manual used for crucifixion of those accused of being witches. It was first published in Germany in 1487 by a Catholic Clergyman, named Heinrich Kramer.

Dating back even further, the first person to be prosecuted was a woman named Hypatia. Hypatia was a Hellenistic Neoplatonist philosopher; astronomer and mathematician in Alexandria trained in Egypt. Her knowledge of astronomy and mathematics led to suspicions of sorcery and subsequently she was beaten by a mob of monks and murdered in 415 A.D.

Heinrich Kramer wrote the Malleus Maleficarum following his expulsion from Innsbruck by the local bishop, the goal extermination.

Ultimately Kramer failed in his attempt to obtain endorsement for this work from top theologians of the Inquisition at the Faculty of Cologne. The book was condemned for recommending unethical and illegal procedures that were inconsistent with Catholic procedures on demonology.

Later Kramer was discredited due to charges of illegal behavior himself. The foundation for the book was Kramer's obsession with the sexual habits of one of the people accused of witchcraft, named Helena Scheuberin. In reality the Malleus Maleficarum was a cross examination manual used as a guideline to extract false confessions from people accused of witchcraft and sorcery.

The Malleus Maleficarum evolved as a foundation for a systematic attempt to find techniques to get people to report that they were things they were not and more importantly, to do things they would not normally do. You must admit this is a perfect recipe for destruction of any life after false accusations and set-ups.

The Malleus Maleficarum historically is documented to be one of the first primitive steps towards behavior modification used as an offensive weapon against an unwitting subject focused on changing a person or the beliefs of people outside of their awareness. The objective is for the subject to fulfill the request of the programmer and to do so without knowing that this is what they are actually doing while oblivious to the motivation of the motivator or Handler.

You must agree that for this type of advanced, psychological warfare and manipulation and influence of the human mind, requires an understanding of the principles of either thought or thought influence, control of the nervous system, the capability of the electromagnetic spectrum and a thorough awareness of the bioelectric energy around the Earth. It requires an understanding of various states of Alpha, our conscious thinking, Beta, Delta, Gamma and Theta the realm of our subconscious mind.

Your dominant brainwave during challenging emotional times will determine your current state of mind. The goal for humanity of which the programmers are aware is too curtail evolution of the Gamma state. Gamma is the higher mental activity, including perceptions, problem solving, and consciousness. It is the awaking in the Gamma consciousness, of which those evil in high places is frightened of because of its clarity and, in reality, they should be. The veil of deception is lifted.

Psychotronic weapons can effectuate "forced memory blanking and induced erroneous actions" as mentioned as a patented frequency manipulation of the human nervous system.

The reading and broadcasting of thoughts, Synthetic Telepathy, etc., "forced manipulation of airways, including externally controlled forced speech," itching in "hard-to-reach areas" and dream control are also, just a few examples of advanced technological, patented capabilities in use today across the board. And the technology is being used in modernized witch hunts of Human Rights Advocates, political dissidents, targeted human guinea pigs, whistleblowers and activist.

"Remote Neural Monitoring is based on the scientific understanding of the interaction of electromagnetics with plasma. Charged particles in the human body can be described in terms of cold plasma and will modify the phase of return signal of a radar pulse," writes an Australian activist, at crazzfiles.com.

"With cm and mm accuracy, a radar return signal will convey information about the electrical state of the section of the body it interacted with. In this respect, the return signal captures an Electroencephalography (EEG) recording of a small area and a spectrograph snapshot. Through progressive scanning of the entire brain in mm sized chunks, an extremely detailed picture of the electrical activity of the human brain can be obtained."

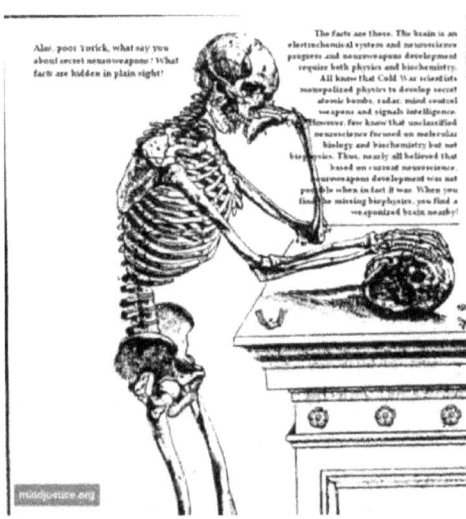

"This is the uniqueness of the human brainwave pattern and why it is referred to as a brain fingerprint. Just like with other brain scanning techniques, this information can be used to decode the subjective experiences of a human (i.e. thoughts, feeling, hearing, touch, body movement, etc.)"

The article further states:

"Radio frequency pulses can be directed with mm accuracy to drive brain or motor neuron activity. This can be used to generate hallucinations and/or move body parts and is the current focus of an unlawful human experimentation program."

The Matrix' is a multi-purpose psychological operation derived from nuclear simulation software. Its purpose is to control human behavior to such a degree that it permits modern computers to predict the future. That is, the solution which has adapted the principle of relativity and particle motions, to humans, interactions and perspectives which can then be projected into the future employing Monte Carlo simulations."

"Monte Carlo simulations come as close as scientist can get to psychic predictions of compliance or noncompliance when applied to individuals, groups, communities and large populations. Monte Carlo stimulation performs risk analysis by building models of possible results by substituting a range of values, a probability distribution, for any factor that has uncertainty."

When I understood this, another dot was connected. I have always felt that those involved in this program seem too know in advance who would become the nonconformist troublemakers of the future likely due to defiance at various stages experimentation. The failure could result in intense focus on thousands of specific individuals forever and intense determination for hopeful technological conquering.

A loving nature and genuine Spirit, I also believe can become a major focus in these operations, for behavior modification. History has proven by assassinations of honorable leaders that good intent for humanity is simply not welcome on this planet.

If many are under influence targeting since birth, these numerous, documented, operations have watched failure, after failure with many to include the heinous manipulative trauma-based tests achieved by subliminal influence for destructive behavior. Manipulated separation from families and friends, beamed creation of high-tech chaos, and tampering is a powerful form of emotional trauma.

Introduction to addicting drugs by intent, promotion of sexual promiscuity, or programmed racism and hatred are all key areas which can be motivated by an outside source. This in and of itself is typical of the power structures awareness of the effectiveness of the psychological effects of tried, tested and proven, methods designed to divide and conquer the humanity family.

In the focus on women, the fact is, men are actually more prone to sexual promiscuity than women. Sexual arousal with women is documented to be mental. This is because women do not have an outside organ attached to our bodies with a separate head of its own. This is why the search for Viagra for women has failed.

With men, the penis can be stimulated as an organ isolated and separate from conscious thought, for example, by sexual stimulated imagery.

As I continued to look back, there were many questionable things around me which I tried to find a source other than what was presented with no valid answer.

I remember always questioning a specific aberration that caught my attention and evoked curiosity, one that stands out in my mind in childhood as well as today. This is people staring off into space with a blank stare on their face, briefly, witnessed many times. It is as if in some type of temporary trance before snapping back to reality. It appears once back in reality, no awareness of the brief trance. It seemed surreal to me. I learned that I am not alone in the observance. This is a question many have sought answer to as well. I wondered if there is any connection with mind invasive technology, brainwashing and mind control.

The question is:

"Why do we "zone out" or stare into space?"

Online Answer:

A small amount of stimulation, such as heavy traffic, can put some people into a dissociative state. An individual in such a state may stare ahead blankly without processing what he sees. To achieve a dissociative state, the body releases opiates into the brain.

The zoning out puts you in some form of a Default Mode Network. During this time your brain is probably processing information and making new connections from a broader perspective based on a recent input. Think of this like when you want to think something over and "sleep on it" and wake up with new insights, except it's happening while you are awake in a smaller form. DMN will maximize the amount your mind can wander and process the information you have just learned.

Today, I recognize this happening with me, albeit rarely, the key with me is my knowing that I am in this program. I have observed that the "zoning out" is always during an out of character negative thought. Of course, both negative and positive thoughts pass through the mind daily, harmlessly.

Today software solution attempts to guide humans to predefined courses of actions, thus generating events similar to the fictionalize tale of robots in the movie and series Westworld.

It does this through a mixture of approaches whilst Remote Neural Monitoring continually improves its ability to control humans by trial and error, which again makes specific human beings useful as focused human guinea pigs targets for decades and the focus, the result of their naturally challenging the system. Their system failure, then becomes vital for an understanding of why and system defects. If you ask me, there is no great mystery. System failure involves vibrating in the highest frequency of love, empathy, and compassion and good intent.

The operatives of this program have long been conquered, programmed and are comfortable in their roles and positions. If there was any deviation and awakening after their intense programming of blind malicious righteousness, allowing them to do what they do and commit heinous acts, they likely would be fired, then even targeted and destroyed also.

Or perhaps, explaining how they can do what they do, without a care result is simply cowardice, or self-serving minds and its inability to access the depth of the machination of this program and the programming as many targeted have. The truth of what it really is, and their acceptance, borders on their insanity.

Through relentless efforts to convince the target that a set-up was the target's own thoughts, the Malleus Maleficarum is played out.

The operatives employed in these positions are paid top dollars and given rank, and more importantly pumped up by self-importance, and narcissism which is also a form of mind control, who in some cases, become challenged by the capabilities of others of whom they are targeting.

Behavior modification is not, nor has ever been about watching targets or specifically targeted populations prosper or progress. It has been in fact congruent with the goals of stifling growth and development.

Psychological Operations in the context of warfare is typically defined as any action which has as its intent the goal of altering or reaffirming the psychology of an enemy and for the benefit of the attacker. Typically, this is often presented in the form of fabricated information that the enemy ultimately acts upon and hopes to convince the target is factual.

However, you cannot change a person's opinion on Weather Modification is if he or she does not first know that weather modification is or is possible.

So, they make some assumptions about targets and given that most attacks are on large populations. These operations are working towards a goal of statistical effectiveness. When working with large populations, these operations are at the same time mitigating the effects of individual agencies in place.

The mechanism of action reveals that people are hardwired at a genetic level to have particular views and that synaptic plasticity can only go so far in changing our views. After all, our views are unique to our environmental experiences and the product of.

Certainly, it should be obvious in the maze of life that certain routes would be inexpressible as a direction for us, if we do not understand the formulation of certain views, ideas, concepts and perceptions which can never hold our understanding because it is not our reality.

Does this make one human being better than another? I think not because people learn various things, in the journey of life at different ages and through various stages of development, and different times yet all valid and valuable to the advancement of the Soul and grand scheme.

Polarity, thus results in the social, political and military issues and an acceptance that manifest, which are not really problems to be resolved, but simply a result of how we function through acceptance of the outside source input.

Any attempt to reshape the world in any other way is doomed to failure, due to the complete ignorance or lack of awareness of this simple fact. Division from unevolved concepts of growth and mental and physical schism against spiritual advancement must always re-emerge as part of the evolutionary process, or we cease to evolve and adapt to our environment. At the foundation of human evolvement, there must be challenges, or experiencing darkness which ultimately creates light or if given lemons learn to make lemonade.

The most powerful method of influence and model is applied to our world in the form of media scripted events, government and military actions.

The purpose is to maintain control, shape human behavior and perception through social conditioning and increasingly direct manipulation of the human brain by Remote Neural

Monitoring, satellites, systems and devices, today psychotronic weaponized drones, and ultimately interference with Divide Order and the natural evolutionary process of consciousness.

The control grid is thus the execution of a computer generated, Brain Interface Computers (BCI) Artificial Intelligence (AI) systems and an apparent strategic long-term game plan evolving as the foundation of the current global order fictionally known as the Matrix resulting in questioning reality. Do you actually think that we are living in a gigantic computer simulation like the one in the Matrix?

Brainwashing is a much more scientific process these days.

Gone are the days of placing someone in a room and playing music, video and employing sleep deprivation alone or the antiquated Lida Machine. Today, it's all about interfacing with the brain and shaping neural circuitry piece-by-piece over the course of a lifetime. With this understanding, it would make perfect sense for studies focused on children pivotal to life assessment.

This is easy for globalist when they control the cultural content and our thoughts through movies, television, news reports, etc., as well as these medium's powerful subliminal hypnotic effect and influence to our subconscious. It's all about learning efficient methods of population control, and the continued control of the "Hive Mind" cybernetics and through methods to accomplish this on shorter-timescales, and by the Artificial Intelligence rise of machines.

It is about creating shock to the brain in the form of controlled deep-seated emotional suffering to alter the personality. With this understanding, it become understandable and obvious how Remote Neural Monitoring has been caught tampering with sexuality as a powerful influencing source.

CHAPTER SIX
Life Long Human Testing

By the age of twelve, I was developing into a young beauty with what many called a movie star figure.

As I mentioned with Richard when he existed the vehicle that day, he asked me about Synthetic Telepathy, and again, thank God I had a logical answer for him that could be traced to patents, patented officially and a technological capability.

I must admit that in this Chapter, I was concerned about admitting any incident that happened around age twelve and on. I realized that reporting the truth could be used as a weapon against me, which this operation had made their main objective and of which many targets have encountered by as stated labeling of mental illness and destruction of their lives after which.

If you connect things that happened from age fifteen to age twenty-five as an anomaly, I realized a connection could be made by what some consider a subjugated quack profession and their Bible, or as some call it, the psychiatric book of lies and fabrications. The DSM VI could be used as a foundation for possible early stages of mental illness, then, this operation would have succeeded by using their most powerful ally. However, the fact is all my life from youngster to young adult, and age twenty-one, before I joined the military, I was never diagnosed with anything even remotely related to or resembling any type of mental illness nor would anyone around me say I acted crazy in any way as if I did at all. I was a normal young woman living a normal life with teenage mistakes, and I felt having normal experiences in life for growth and maturity. However, now when I look back there were many instances during which time I always felt that something just did not seem right as if there was a subtle alternate reality at play. I don't think it was just me. Based on what I have learned many targeted make the same claim.

Admittedly, under the targeting, I like thousands, actually wondered if something was truly wrong with me, and whether I was losing my mind. It surely felt like it.

I met my ex-husband and we were married a year after going into the military and married for thirteen years, three children. A lot of the things that happened growing up I brushed them off, or forgot about them until the targeting style of today awakened memories.

As I continued the search for understanding and awareness of the capability of an outside source for substantiation, one thing was certain, the official patents, hundreds are not sitting on a shelf somewhere collecting. At first, I did not understand how things that were personal experiences in childhood were being played back no officially in this program. There was also a continual attack on how I look and a ridiculous statement verbalized by the goons using acoustic through the wall radar technology as well by this operation repeating "She thinks she's cute."

Where did I remember this from, now years and years later as an adult? Oh yeah, it originated around the age 13 as a source of confusion for me and hurt feelings.

With this simple statement, and it repeated by the goon squad yet another connect was made to the past and a long time ago.

I had no choice but to wonder had today's operation dug into my profile and uncovered something which hurt my 'little' feelings as a young teenager and determined it useful in the now intense psychological operation around me?

Believe or not, if you understand this program, it appears they had done just that in a desperate search for any weakness to exploit and hope to break me. How else would they have known how hurt I was as a thirteen-year old when people kept saying this about me, as I innocently blossomed into becoming a lovely young teenager to destroy my self-esteem?

To drive home the capability of the technological capability of suggestibility, this also happened, with my now fully aware of tracking, monitoring and powerful mind invasive technology used around me today.

A few years ago, also documented in one of the other books, while the effort around me intensified, after completing the first five books in this series, the realization of beamed suggestion was brought home by use of perfect strangers, who likely had no idea where negative thoughts about me came from or more importantly, what prompted them to repeat the subliminal influence, and what is called "Direct-to-Speech."

The concept of Direct-to-Speech can be referenced to demonstrate how an incoming signal can instruct your bioelectric brainwaves which then can be used to interrupt normal speech by essentially cutting off the vocal speech pattern from your brain thought pattern, and inserting in real-time a new vocal pattern; which could also result in a person speaking gibberish without control. There are several YouTube videos of this happening to news reporters.

While standing in a check-out line a complete stranger uttered, "She thinks she's cute."

When I turned to look her in the eyes, she also had that same zoned out look on her face that registered that she probably did not have a clue where the suggestion or her verbalizing it to a complete stranger came from or that it even came out of her mouth.

After this operation had beamed cooked a large bald spot off the top of my hair, which also happens too many targeted, about the size of a small orange, I tried to cover it one day as I ran errand, by styling my hair differently. While standing in a check-out line, this time a young man and young woman got behind me, after I witnessed them maneuvering to get as near to me as possible. When they finally did, then they reported loudly that I had a bald spot. I just turned and looked at them and thought whatever in full awareness of influence. And, on top of this these two looked like USAF military personnel from a nearby major air base about 30 miles from my home.

I mean really... Who does this to a complete stranger?

One thing was striking, after this silly comment said now to a full-grown adult, with grandchildren, my mood would immediately drastically drop, and a feeling of depression engulfed me after the silly "She thinks she's cute" remark. The problem being that I was long past thirteen years old and mentally could care less, yet I was engulfed with a depression frequency.

I had to consider, based on the impact that around thirteen, this operation became aware of the revelation this comment had on a young girl and the negative emotion likely had been cloned and stored, and now was being played back years later. Those at the helm of the operation were hoping to ignite, the sadness of a young girl, by creating synthetic depression done by then repeating this over and over again on a daily basis and attempting to drop create depression, believe it or not. It was bizarre to say the least that my mood was immediately altered by this pure foolishness and physically affected by something that did not matter to me at all. At that age perhaps, I desperately wanting to be liked as most do but today I could care less.

These operations can beam people to become jealous of you focused on how you look or your car, material objects, etc., to include your accomplishments even when they are not jealous at all. Negative influence is highly effective with those they hope to use by, likely first highlighting their failure to reach their life goals. This program is no joke and highly perfected and I say, reluctantly, brilliantly structured.

At first, they were saying the exact same thing from thirteen years old. Then they switched to repeating, their version of "She ain't that cute," over and over again. What a waste of time, energy and resources of these immature silly people using this technology.

At a certain age, you begin to care less about what people think of you and more about what you think of yourself. This is the end result of conquering battle after battle within the maze of life, time and time again. It becomes excellent self-esteem building.

One summer, I returned to my surrogate grandmothers, at age 14 to attend summer school classes. Because of hypersensitivity, and hurt feeling, I did not want to go to school, in what some would call today a tactic known by thousands in this program as Nazi originated Mobbing at school key in this program for decades.

Always well groomed, now at Mother Julia's house for the summer, a neighborhood girl would sit on her porch every morning watching me like a hawk going and coming from school leaving around 8:00 a.m. and returning around noon.

One day while inside, and another relative visiting, I heard a commotion outside. In the street were two sisters fighting, one of whom was the girl with a twisted look on her face of someone appearing to need an Exorcism focused on me each day.

I looked out the window at them, but because I could not hear what was being said, or why they were fighting, then decided to go on the porch, so I opened the door and stepped out. As soon as I did, the girl stopped fighting her little sister which was right in front of my house, and the full of her attention and anger was immediately redirected to me.

"What are you looking at?" she said eyes glaring, narrowed with unjustified hatred.

Shocked at how quickly her attention was immediately refocused on me, I said nothing.

"The next time I see you I am going to do the same thing to you" she said. "I bet you will run and are scared now to come off the porch." "And what are you doing here anyway?" she said. "You don't belong in this neighborhood." I was there because Mother Julia owned four houses there. She said it as if I was too good for the neighborhood.

To this day, I do not know what possessed me to get off my porch in a fenced yard and walk over to her, but perhaps it was likely to prove I was not scared. I was not angry, and had no ill feeling for her at all but yet there I went after the challenge.

From that point on, this girl began to whip me like there was no tomorrow, she bit me, and while on top of me pounded my head into the concrete to the point of if, she continued, a possible concussion.

Later after the dust settled, I recalled, and will never forget, that during the fight, what I now realize was possible subliminal influence of her to act. I say this because I am not psychic, but knew what she was going to do. This was to scratch, and scar my face.

As if on command, while on top of me she briefly stared blankly into the sky, zoning out for a few seconds, then as if having gotten her instructions, she took both of her hands using her fingernails and started at the top of my forehead and raked all ten fingernails down to the bottom of my cheeks, leaving a trail of bleeding fingernail marks in a Zebra pattern also from ear to ear.

After mission accomplished, she jumped up, appearing satisfied and I stood bewildered. Everything had happened so fast. I wondered why I knew beforehand what she was going to do which seemed odd to me and it appeared, again almost if she had been given instructions.

By this time, the visiting Aunt had run outside and put a large butcher knife in my hand telling me to stab her. Something within me would not allow me to do this. I dropped the knife in the street. When the girl saw me do this, her and her sister retreated to her house.

The fact is had I used the knife my life could have been destroyed and it would have set my path in a totally different direction resulting from likely juvenile incarceration, for attempted murder, detained in a juvenile detention center, and some would argue one of many human lab rat facilities and programming sites for youngsters.

When I think back about this incident, in full recollection today, I realized the 'possibility' that those these operations have been unsuccessful with are those who have become "primary targets" today. We are the loose pegs of all races under continuous focus to break and proven a major challenge to control.

In 2014 while visiting and playing with my granddaughter in another state, she said, "I can't say whore about my Nana" out of the clear blue. With her, I had also witnessed many zone-out stares, and because of this, and many things once leaving me in lost in confusion, the once unexplainable was now linked to the possibility.

As my mind began to flutter back in time, later, as the oldest girl blossom into a real beauty there were some oddity around her as well. When she later became officially targeted, to use against me, I wondered if this operation may have watched her growing, before becoming official around her for use as a pawn, and decided her useful for sexual victimization along with thousands upon thousands of girls all over the world.

Human Trafficking is widespread today and documented to involve high level professionals, doctors, lawyers, judges, policemen, politicians, etc. And based on what I had learned resulting from my recognition of what was happening around many, it is not farfetched.

We monitored everything our children watched on TV and my granddaughter's mother, the second oldest, believing she was going to die around age six, from some unknown cause, out of the blue admittedly struck me as odd. What was stranger was the intensity of her belief. Could it have been influenced I now wondered? Recently, she asked me did I remember her feeling this way as a child confused herself. The only thing I could tell her, now fully awakened it that it was likely part of this program of very likely generational targeting by individuals who work these programs of scientific Remote Neural Monitoring studies on a massive scale from their time of hiring until retirement and someone else hired then fills the position and carries on from originally nationwide military bases.

Today, threats of harming my daughters and grandchildren abound, in the background of my phone while talking to them. And, each visit has consistently been a major attempt at high-tech chaos.

I have been asked several times to stop what I am doing and this operation will leave me alone or else. But I had come too far to turn back now. What am I supposed to do, discontinue the books series, blog, website, docudrama, because they are demanding that I do?

Once I was forced to leave abruptly before the planned flight, during a visit also to the second daughter's house. I woke unable to sleep with my legs intensely tortured by the energy weapon. My legs were always a major focus when out of state because the inability to walk could ruin any vacation. Unable to sleep, as my body vibrated to the irrigation of likely a drone, sleeping in the same bed as my daughter and granddaughter, I watched as my daughter woke and appeared to be sleep-walking.

I watched her go downstairs, open the drawer with silverware and knives in it and then back upstairs still in a trance like state. When I asked her what was wrong, she hissed at me with an evil look in her eyes. She then got back in bed and went to sleep. What if they were trying to program her to harm me? If so love saved me.

Anything is possible today and the objective is to use others for no accountability and no one believes it is mind-invasive technology programming and the perfect official crime.

The next morning, she was extremely angry with me, out of the blue and a major argument evolved over nothing. I left because, the objective of this operation had several times created agitation in my environment, especially with my children and from past experience, it led to this operation's hopeful police interaction while in another state and strategic confirmation of my insanity, by use of my children for Episode 999. When I saw where it was going, and assessed the situation, I high-tailed it out of there with a quickness not wanting to take any chances, and returned home.

During two separate visits, I witnessed, with my own two eyes, FBI agents going into the condo next door, and following me when I went out to get food, obviously alerted to my visit beforehand, and also alerting the community, when she lived, then at another location in this nationwide massive targeting program of high-tech monitoring from state to state and country to country. And that night, I could feel the portable technology placed against the wall and hear it being positioned.

Recently, as I worked on this book, she asked me to visit again. I thought heck no and especially not until this book was finished.

I was under a major focus as I worked and covert efforts intensified around me. In fact, this manuscript has been tampered with so much, to the point of an earlier version being deleted from my computer and unable to retrieve. Research revealed that tampering can be done by the radio frequency technology they use to also hack a computer and that disconnecting my MacBook Pro from the internet did not help as Apple Tech Support has suggested. When I Goggled can a computer be hacked if not connected to the internet, the article excerpted earlier revealed that it could be and definitely by Intel agencies. However, the proof of cleverly inserted errors in grammar and punctuation was all I needed after proofreading over and over again, and everything I do monitored without exception. I finally hoped that the information in this book would prevail as important and surpass the official tampering. The hope is to portray writers of this subject as illiterate and hope the public is turned off by the tampering.

I believe it had been compromised when I had it fixed and left it with a computer repairer for three days. As a result, I had to dust off the old computer to proofread. If left as it was, it was trash. I continued to make corrections. And, they were doing it so cleverly that even spell check would not catch mistakes. For example, a word like 'been' was changed to Ben and many other words changed in the exact same manner, their, became there, etc., of which spell check would not correct because they were spelled correctly.

Getting a publisher was out of the question, I learned from experiences with the other books due to intimidation by these agencies involved in this program not wanting exposure, much less hiring a proofreader and editor for them to intimidate.

Nearly all of these books are published on Amazon's subsidiary or which many authors have used or what are called Vanity Publishing companies. This is where they pay a small fee for full publishing services different from traditional publishing which give a royalty advance, these two methods are how 90% of the boons on this specific topic are published.

They were also trying to convince me that my granddaughter's Hispanic father was a pedophile and molesting her when around them. And when I thought about it, this may even have been what they were using to upset my daughter that night while sleep. They even went so far as to have the child say things to me that would formulate the negative thought of a reality which was obviously wrong. I had already scared him half to death with mentioning any possibility and he did not want to tangle with me nor my policeman ex-husband.

This is the nature of this official high-tech, COINTELPRO, psychological warfare operation.

My mentioning it to him, of which, he was rightfully insulted resulted in him never forgiving me. My doing this was founded purely on the understanding of this program's inner workings and an understanding that anyone could be used and abused without a care in the world. Yes, creating sexual deviances is vital and high on the list in the trap to enslave or use anyone, no matter who you are or what position you hold or may have in life.

Right before the demise of my marriage, I had a very odd one dream night. I woke feeling high although I had tried drugs rarely before going into the military. It was now twenty years later and my marriage was on the rocks. The fact is drug dreams are documented to be a major cause for enticing a person to use drugs. Get the picture?

Do these operations know this, after years of drug infiltration, and ongoing illegal and legal drug testing today and yesterday? Absolutely! Drugs and influence to use drugs continues to play a major role in mind control testing programs?

It is the subconscious mind of which this program focuses on to include while the target sleeps using electromagnetic technology hoping to tamper with consciousness by frequency influence and creating thoughts which can be useful even in the future for the subject's demise.

Usually when the dreams are so realistic that the person wakes up and feels like they were factually intoxicated, key is the feeling left with the person of temptation or even possibility.

I now question why my Mother Julia kept Sage bunches tied around the front door while growing up. Sage is regarded as a way to ward off evil spirits. Although this woman loved and adored me, I frequency saw her too with saw the same blank stare trance.

After a typical teenage argument with her, I recalled that one day, while sitting on the top of the stairs at her home, I listened to her talking with a visiting disgusting looking relative of hers visiting from the South who to me appeared to be mentally deranged. I sat enthralled because the conversation was about me. When I heard him say he was going to get me, I flew into a panic. When I saw him get up and head towards my direction peeking around the corner, I raced upstairs, and locked the door and placed boxes, chairs, and anything I could push against the door.

This was yet another unexplained event, of which she appeared to co-sign him hurting or possibly raping me. It made absolutely no sense at all and today, people that knew her report she loved me dearly and more than life itself. However, today, I must consider the dynamic of this program and ongoing trauma-based mind control and rape. Had this buffoon raped me my mind and heart would have been shattered and broken. This especially true if it appeared to be set-up from someone I deeply love and who had protected and nurtured me all of my life and then just eleven years old.

One thing is certain, she definitely felt evil around us, but unlike many then or today had no clue of the possibility of the focus of a vicious scientific experimentation program.

CHAPTER SEVEN

Eyes Wide Open

"**H**ow did she get so far" he said, beamed from the operation center overseeing the drone positioned over my home through the ceiling. He obviously meant books, etc.

The response, "No one expected it."

If you want a trail of this program, the trail starts with the most publicized mind control study, MKULTRA up to present day with others studies just as significant in the Project Paperclip conspiracy, which included Projects, Bluebird, Artichoke and many others.

There were few victories MKULTRA through a "civil suit," in which no criminal charges were able to be brought, however produced a presidential apology, compensations for victims scarred for life due to physical and mental torturing and abuse from the many different experimentations and tactics used in these projects.

The Defense Advanced Research Projects Agency (DARPA) is an agency of the United States Department of Defense. This agency is responsible for ongoing technologies for use by the military which trickle down to federal, state and local police departments. With DARPA neurological research and development of cybernetics to bring autonomy into Artificial Intelligence continues. These are programmed AI machines that study human behavior to collect empirical data and learn from themselves.

The psychological burden of targeting is so severe that it has drove some to the point of suicide. I do not see the win in any of this except a satisfaction of monstrous evil. Currently these same and other tactics are still being used proven effective resulting from ongoing programs. See torture memos in Wikipedia.

15,000 pages of declassified documentation on non-consensual human experimentation in the United States are available for confirmation. Yes, blacks were enslaved because of the color of their skin, in a Nazi mentality but today, everyone is considered black and probably always has been.

In 1990, over 1,500 six-month old Los Angeles black and Hispanic babies were given an experimental measles vaccine, never informing parents of the potential harm revealing ongoing testing on poor communities.

Without a doubt, an outside influence, has sought in the very beginning, and every way and continues to seek to maintain control by engineered chaos on a much larger scale. Without this type of created confusion, the Controller could not thrive and continue to survive to the present day.

The ego is very powerful and operates in the realm of self-preservation. Ego when used for the good of humanity is a spark of Deity. However, it is not if not being used for good. We are here to learn to love; however, evil believes we are here to instead learn hate.

The mind has its seat in and over the brain. The spirit has its seat in the Divine Ego, or God- spark, which has its seat in the spirit.

In two excerpts from Chapter V: The Constitution of Man – Sacred Texts, it is written: The constitution of man is therefore in essence a Spark of the divine Fire....

It further reads:

> "The ego is the man during the human state of evolution; he is the nearest correspondence, in fact, to the ordinary unscientific conception of the soul. He lives unchanged (except for his growth) from the moment of individualization until humanity is transcended and merged into divinity. He is in no way affected by what we call birth and death; what we commonly consider as his life is only a day in life. The body which we see, the body which is born and dies, is a garment which he put on for the purposes of a certain part of his evolution."

I began walking a path founded on faith, trusting and believing that the position I had been place into had ultimately good meaning and the experiences and suffering were not in vain. "All things are working together for God for those who love God" became my motto. It is my hope with many others to assist by exposure of the hidden evil.

Obviously, there are those who do not share this belief and are Hell bent, in this polar existence, of good and evil or checkered existence who, endeavor to create a whole different reality. This is proven without a doubt by the bombardment of the decisive Luciferian agenda unfolding everywhere we turn and as stated sex playing a profound role in the scheme.

Sexual physical arousal is considered by occultists to be extremely potent and can be channeled for magical purposes and was factually a practice used by Aleister Crowley. Magic, in the context of Aliester Crowley's Thelema, is a term used to differentiate the occult from stage, theatrical magic and is defined as "the Science and Art of causing change to occur in conformity with will," including both "mundane" acts of will as well as ritual magic or more definitive electromagnetism mental and physical control.

Crowley, a student of ritual magic wrote that "it is theoretically possible to cause in any object any change of which that object is capable by nature." In other words, both the negative, which typically lays dormant and positive within, can be manipulated.

Little known to the average individual, Aleister Crowley was at the heart of one of the most influential movements of the 20th and 21st centuries through a secret society he founded and named the Ordo Templi Orientis (OTO).

He was very highly regarded and connected to the British M15, the British equivalent of the United States' Federal Bureau of Investigation with the British M16 said to be comparable to the United States Central Intelligence Agency. Aliester Crowley's understanding of occult magic and practices also, connected and involved him with the goals of the Global Elite.

The OTO is today's best-known of the hard-core, British-based Satanist cults. Like the Lucis Trust, the OTO is a direct off-shoot of the work of Britain's leading twentieth-century Satanist, Theosophy leader Aliester Crowley. OTO enthusiasts claim this organization is an offshoot of

Templar freemasonry, and hint at very influential protection from among Templars very high in British freemasonry and leaders of the New World Order.

Sex magic, a very powerful energy or frequency, is the use of the sex act, or the energies, passions or arousals it evokes, as a point of which to focus the will or magical desire in the non- sexual world. Basically, sex magic can be used by the occultist to harness the emitted frequency of sexual energy then redirect it somewhere useful for their empowerment as if packaged. If, the sexual energy is gained from the result of pain, sorrow, victimization or abuse, it also carries a higher level of frequency derived from the sorrow. There is a powerful secret force that lurks within sexual desire, and the act itself can be used to enhance life creatively, or to destroy it pathologically.

Creation of rape, pedophilia, to include bestiality, etc., all carry a specific type of energy by duality which feeds the same frequency energy and creates powerfully different results. Understand that the programmers are serious about the relentless effort to capture, entrap and maintain enslavement over the human mind and thought and it is as revealed earlier a subtle influence and always opening the door by being hideous.

Many will never forget the jaw-dropping episode of a segment of "Nip Tuck" in which a woman, Shari Noble had been applying peanut butter between her legs for her dog to lick it clean shown on the Global elite-controlled media.

The husband, military, arrives home from Iraq and visits Shari at the hospital, and tells Shari Noble he found the open jar of peanut butter by their bed and that he knows she uses it to seduce the dog. Shari is at the hospital to have reconstructive surgery for her breast nipple after her dog, a pit-bull mix, has bitten it off before he arrives.

One would ask, why would this disgust be publicized on national television? One thing is certain it lingers in the mind of some. But more importantly, could it have any influence as acceptable, by the intentional gradual indoctrination of acceptability through television? This type of subliminal programming is not a farfetched concept.

"Sex and the City" drew the line at Bestiality content writes a June 2018 article in the New York Post.

The incident in question happened during the series one episode titled "The Monogamist." The scene unfolded between Charlotte (Kristin Davis) and Michael, a guy she had just started dating kept pressuring her to perform oral sex on him. Charlotte agitated storms out saying he only wants her for this.

What made the cutting room floor was the original scene in which Charlotte returns sees that Michael's golden retriever was going down on him. Initially, before the cut, he was shot putting peanut butter on his penis for the dog also for the dog to lick before Charlotte walked back in.

The great power of sex lies, again, in the powerful creative force that brings forth life itself which is being the honed for use by Devil's spawns.

Sex, is the essence of creation; from the birth of all life, to the birth of passions to the birth of creativity as a powerful, positive driving force which is being intentionally tampered with. For centuries man, for good or evil, has tried to channel this energy into more fulfilling areas and higher states of consciousness.

People have debated the various mental and physical benefits and drawbacks of both sexual abstinence and sexual activity. Purported benefits of abstinence bring mental energy, sensitivity, compassion, empathy, creativity, and advanced spirituality. Harnessing sexual energy and sexual transmutation of the energy into high ideals can send productivity into overdrive in all areas of life itself. The results of many well-known greats who practiced abstinence reveal that it elevates creativity such as Gandhi, Leonardo Da Vinci and even Nikola Tesla.

The reality is that occultist knows what has been long known as well as a powerful tool. The intensity of sexual energy cannot be matched with any emotion known to humans. The force instigated by sexual energy can override even the most intense fears in a human and there is enough evidence to suggest that humans can undertake some extremely risky behaviors under the influence of this mesmerizing energy, which they would not dream of doing if they were not taken up by it. No other force has the power to inject the amount of courage, fearlessness, imagination, impetus, motivation and creativity that the force of sexual energy has.

This is one reason why sex is the most sought-after activity by human beings and why sex is the major topic of discussion in all spiritual and religious literatures and promulgated intensely today everywhere and by every means. The fact is, the key to the power of sexual energy, and its usefulness to the occultist, can be seen in captivating humanity by it, by capturing the power of thought, resulting in over indulgence and creating unbalanced behavior and is typically documented to be found in humans who fail to develop the conscious channeling of this energy into their creative expression, personal power, well-being and enlightened state of mind.

While there are those seeking enlightenment through the proper channeling and use of the Kundalini life force energy, for centuries occultist have successfully channeled the energy, electromagnetically, into lower consciousness states and blasted humanity with various, captivating techniques for sexual mind control entrainment.

As reiterated in this book, obviously, the objective is to keep humanity operating closer to the animal kingdom of lust, physical gratification instead of spiritual gratification. The goal is to block the creative process connected to our Divine Creator and the creative force of good.

This is a magical key and the reason for the prolific bombardment and influence via the Globalist controlled entertainment industry and promoted by both music and television, and sexual promotion everywhere in movies, and even basic family programs geared towards children.

Walt Disney, one of the 13 Globalist families, who are reportedly have been heavily involved in sexual entrainment of children clandestinely with not only Alice in Wonderland production, but may others reportedly.

Apart from the Tantric and Taoist sexual energy practices, there are countless other references to the hidden powers within sex. The Kabbalah for instance, sees sexual desire as the deepest spiritual expression one can have. And in Yogic philosophy, our pure sexual energy is dormant until awakened in its highest form the "Kundalini," but only if used for good.

Many esoteric groups also practice something called" Sex Magic," which can be connected to Psychologist Carl Jung's notion of sexual alchemy (similar to Freud's sexual sublimation. Sublimation is probably the most useful and constructive of the defense mechanisms as it takes the energy of something that is potentially harmful and turns it into something good and useful.

Freud believed that the greatest achievements in civilization were due to the effective sublimation of our sexual and aggressive urges that are sourced in the Id and then channeled by the Ego as directed by the Super ego. Sexual alchemy essentially says that with pure intense will power we can transform the raw energy from our libidos into golden creativity and the same truth is likely regarding transformation of negative sexual energy as well for evil purpose.

The key to the targeting and manipulation of both children and adults by sexual distortion the energy and sexual identify in the early stages of development, can be attributed to ill use of this knowledge and this type of magic and the recognition of it being equated with the "life force."

Through the ritualistic use of sexual techniques, inspired by Tantric schools of the East, the initiate can use the immense potency of sexual energy to reach higher realms of spirituality or the occultist or minions to reach depravity.

Aleister Crowley proclaimed and embraced the labeling of himself honorably and his doctrine as being that of "The Great Beast 666."

"The order had rediscovered the great secret of the Knights Templar, the magic of sex, not only the key to ancient Egyptian and hermetic tradition, but to all the secrets of nature, all the symbolism of Freemasonry, and all systems of Religion." [11. Theodor Reuss, Oriflamme]

To set in motion the "occult forces which would result in the illumination of all by 2000 A.D.," Crowley became convinced that his mission was to "cure the world from sexual repression." To achieve his goal, he became determined to study every detail of sexual behavior and bring every sexual impulse up to the region of rational consciousness. To this end he experimented with altered states of consciousness, including hashish, cocaine and opium.

Crowley would eventually introduce (not without protest) the practice of homosexual sex magic into the Ordo as one of the highest degrees of the Order for he believed it to be the most powerful formula. [12. Jason Newcomb, Sexual Sorcery]

It is no wonder today that the Illuminati controlled entertainment industry is rumored to be a haven for extreme wealth and inside mansion homosexual parties, pedophilia, orgies and blood sacrifice. Or even the promotion of televised transgender acceptance and desensitizing through award for those likely MKULTA programmed, who after gender transformation is named "Woman of the Year. Reportedly those who sign their name in blood relinquishing their soul for fame and fortune.

CHAPTER EIGHT

Unification Nationwide and Globally

The highly perfected program could not have been developed without input from psychologists, psychiatrists, electronic and aerodynamic engineers, highly trained computer programmers and technicians, probably physicists, chemists, and pharmacists, mind control and interrogation experts, i.e., the intelligence community, and others. Nor could it have been developed without the knowledge of the FBI or the blessings of high-level government alphabet agencies, the military and the Military Industrial complex.

However, above and beyond the call of duty, the Association of Psychiatry continues to play a pivotal role. Psychiatry has expanded their vast control by broadening the definition of mental illness to include political disobedience. Thus, psychiatry became in the past and today a tool of and ally of a government, particularly those of a Nazi-foundation and scientific control regime.

Perhaps the relentless hope to convince me of being a liar would have merited some validity had not the MK ULTRA program continued to sexually aroused me relentlessly against my will with no mental connection to the arousal which is the key.

The downward spiral of unexplained actions can be devastating to the psyche and, again, this was happening before I went to the VA for help. Again, and again, because of this, I knew that I had factually been in this program of subtle influence, for years and while on the facility, I was still being targeted, unknowingly.

My whole life and the lives of many, many others had become part of a massive program of which we were not supposed to figure out and if figured out then have the audacity to publicize, the targeting resulted in then being marked for covert death strategically.

As I woke each morning, the hope that the brainwashing had worked during sleep quickly diminished for the group working on me while I slept. The character of those involved was so low that many wondered if it was factually true what one noteworthy whistleblower and American journalist reported as the character of those involved saying, "these operations would be run by the underclass" for obvious reasons and they themselves operating at a lowered consciousness frequency, and similar also to the "Flying Monkeys" in the Wizard of Oz. They were under orders by the Wicked Witch of the East.

These are those viciously in search of importance. They would be elevated in these positions because of a desperation need to be someone important and would kill to remain in these positions they belief offering them importance and also the source of their livelihood.

It would take an underclass of people from many races to do the work of this evil. Was it because they have very little opportunity to effectively do anything worthwhile or were they victim's themselves of intense programming stealing their soul by their acting out horrifically wrong acts?

These jobs afforded the destruction of the lives of others, while hiding behind advanced technology, unleashing destructive psychopathic alter egos approved by pathological power.

The game is to target children early which eventually leads to a misdiagnosis which then makes them prey for Big Pharma's ongoing mind control drugs, or even illegal drug influence, beginning with children in the early years. Attention Deficit Disorder is a typical and key effort for example.

Again, and again, it must be understood that this technology, is no joke, nor are those programmed to enforce it, who possibly, if military, are for sure heavily indoctrinate themselves sitting at the helm. They operate on a level of ruthlessness and dish out such cruelty that many cannot believe it. In fact, it is too their advantage that the public does not believe that it is happening. Anyone's brain, I repeat, anyone's brain can be hacked, anyone, family members, neighbors, children, everyone.

Former NSA contractor Edward Snowden has revealed the real reason the government use of mass surveillance on the public. He reported it is not to protect from terrorism, or the War on Terror, but to brainwash the public into accepting the use of social conditioning so that control mechanisms can be asserted upon the population and unleashing of the global system all over the Earth itself for the New World Order.

We are in an era in which data processors of the human body and mind may be manipulated or debilitated. And an entirely new arsenal of weapons, based on devices designed to introduce subliminal messages or to alter the body's psychological and data-processing capabilities, can be used to incapacitate individuals.

These weapons, are known as Remote Neural Monitoring, and various others types are aim to control or alter the psyche, or to attack the various sensory and data-processing systems of the human organism. In both cases, the goal is to confuse or destroy the signals that normally keep the body in equilibrium.

Nearly every NATO country in the world is involved with use of advanced technology under the New World Order objective. And, this is especially true of those who are viewed as troublemaking protesters, protesting against advanced technology use and "Ongoing War Inc." GMO food manipulation, Weather Modification, and deadly radiation of all life in the ocean, deadly to humans as slow-kill, after Fukushima, is not reported by the media nor is Big Pharma gain from billion dollar drug profits.

The fact is Remote Neural Monitoring, has been all dressed up and promoted as useful to humanity. This is just half of the truth. Microwave energy weapons can be used to give a target a covert, undetectable, Heart Attack or Stroke or cancer, etc. However, it's main use today as reported by growing numbers, and packaged to the public, is as a method to revolutionize the "War on Terror," and hunt for terrorist. "Thought Police" use crime detection before a crime is committed, and an innocent, passing thought, has destroyed many test subjects actually entrapped. Those used can be then targeted in bogus investigations of which never see the light of day in a Courtroom of which many have tried and tried and tried. Any type of exposure of what is really happening, or the Malleus Maleficarum tactic used is the last thing these operations want publicized.

As a result, a person is literally interrogated, with repetitive with suggestions, even while sleeping to lock information into the subconscious, and it played out inside their heads while awake if it sticks. If you do not accept what the effort is attempting to force feed into your sub-consciousness, operatives then escalate typically, covert microwave Directed Energy Weapon coercive, "Pain Ray" torture using the military Active Denial System in many formats.

The fact is, today supercomputers are sending messages through biometrics, DNA, iris, gait, facial recognition, and fingerprints uploaded of millions into supercomputers which then can track, harass, degrade, manipulate, influence and control a person. This can happen even if deep within a bunker beneath the earth. Biometrics is the new paradigm.

Well over seventy years of neuro-electromagnetic, non-consensual, involuntary human guinea pig use, the end result has bought radio frequency, electromagnetic frequency (EMF) extremely low frequency technology to the forefront in the technocratic state.

If you want to understand the full power of technology at play today, simply look to Hollywood. Ten of the best movies about mind control, and brainwashing worth your time can be Googled.

In the realm of psychological electronic weapons, one thing is certain and must be remembered, is they can record emotional thought, our reactions, but cannot record the feelings which are far more powerful than thought in this respect. Because of this there is hope and this is likely the cause for the system failures with many. They can record, agitation, blood pressure rising, nervousness, heart pounding, intense breathing, and listen to your thoughts when processing these experiences.

Thank God Almighty, there is no technology that can record feelings which come from the heart and operate within the higher electromagnetic love frequency nor do these operations want too. Human intuition & wisdom are also not "machinable." They operate at a higher frequency then what this evil is attuned to.

The thought of love can be recorded but no technology can capture the invisibility of electromagnetic feelings of kindness, caring, compassion, empathy, etc., or more importantly altruistic concern for others and humanity.

And when technology is trying to create the opposite, this is where it becomes an immoral, unethical, horrific and an evil intrusion. Our thoughts, feelings and intellect work together as one setting the stage for our conclusions. This is within the understanding that thoughts and beliefs fluctuate as part of development until a final conclusion is reached as the mind mulls over experiences.

There are some estimates that millions are being remotely monitored at a speed of 20 billion bits per second, via highly advanced super computers, globally.

Remote Neural Monitoring (RNM) combines functioning at different levels, for example Signals Intelligence (SIGINT). SIGINT can remotely detect and monitor a person's bioelectric field. SIGINT has the proprietary ability to monitor remotely and non-invasively, information in the human brain. This is done by digitally decoding the evoked potentials in the 30-50 Hz, by the 5- milliwatt electromagnetic emissions of the brain. This system uses electromagnetic frequencies (EMF), to stimulate the brain for RNM and the electronic brain link (EBL). An invention called the Neurophone, an older technology today lies at the center of the debate as communication is carried to the brain via the nervous system.

Your peripheral nervous system connects the nerves from the brain and spinal cord (central nervous system) to the rest of the body, including the arms, hands, feet, legs, internal organs, mouth and face. The job of these nerves is to deliver signals about physical sensations back to your brain which is how the technology is being used to create pain inside of the body as many target's report from the inside out as well beamed weapons.

Neuropathy is an example of a condition that people with Diabetes experience revealing the power of the nervous system. Nerve damage, neuropathy, leads to amputation, for example, of a foot. This is likely why the beam reportedly can be intensely focused on nerves, due to the understanding of nerve power or other health deteriorating areas for various reasons.

Strategically, any claim of technological damage resulting from specific damage by technology used by government operatives would be written off as similar results of diabetes and it all strategic to keep the truth hidden. Many who make it out of the web of deceit and total entrapment of their minds by awareness, and regain their lives prove to have always been intelligent, caring people who were entrapped.

How, why and what is it really all about?

It is about mind control through emotional domination?

In order to answer the above question, one must be aware of how technological advancement is playing a key and definitive role today in total population.

Here are what other notables have to say:

The wave of negativity created by [the cabal's] frequency machine HAARP, scalar wave technology, demonic wars, vaccinations, poisonous food, drinks, entertainment and smart grid, are all attempts to keep us vibrationally resonating in their matrix game. The more blood they can spill, the more children they can terrorize, the more animals they can slaughter, all add to the lowering of planetary frequency. Their aim is to distract and terrorize us on every level of our being...it's called traumatic mind control. **(Elva Thompson)**

Species-wide traumatic mind control occurs because "the game is not about money; it's about power and it's about energy – or more specifically, human energy. It's about setting up a system where people at the top sit around all day, managing the human farm, while their economic or literal slaves do all the work to benefit the controllers." **(Makia Freeman)**

Hidden knowledge used in hopes of maintain control

"Third dimensional reality is animated fractal geometry. The resonance of the broadcast is decoded and interpreted into five sense reality by our antennae, the chakra system in our bodies. The five-band spectrum defines the perimeters of our physical consciousness, the world of touch, taste, and smell, light and sound. **(Elva Thompson)**

This 5-band spectrum is electromagnetically manipulated by the cabal because the "powerful feel entitled to get away with anything and everything because they've been getting away with anything and everything for decades while the American economy, society and culture all rotted from within." **(Charles Hugh Smith)**

They manage the herd and maintain dominance by creating waves of negative frequency. They fear "we will resonate on the [cosmic] information field available at this time and awaken to their scam. But their greatest fear is that we will claim sovereignty over our own minds, and regain our cosmic consciousness... our 90% de-activated DNA." **(Elva Thompson)**

Nobody said the road would be easy, however, when it is paved with government operations seeking to ensure it remains in power, while carelessly destroying lives of those perceived inferior, the road becomes a life or death challenge that tests your will, inner strength, courage, fortitude, stamina, determination and endurance.

The first thing we must consider is that the police, federal, state, local, and the military were created specifically to control working class and poor people and to do the bidding of a higher-level control mechanism.

CHAPTER NINE

Behold, I Send you Out as Sheep in the Midst of Wolves; So be Wise as Serpents and Innocent as a Doves

S ymbolically, a maze is designed to challenge our direction and choices in life. The purpose of our movement through the maze is to find our true selves and our place in the grand scheme of this existence and our role in the world and hopefully it is to make the world a better place.

> "In general, the greater the understanding, the greater the delusion; the more intelligent, the less sane."
>
> **~ George Orwell "1984"**

The fifties laid the groundwork for a new breed of "brain warfare" and testing efforts became vital aggressed in the sixties on.

In the fifties, brainwashing techniques and information began to not only terrify but also fascinate the American public. As "Prisoners of War" returned home from China, Korea and the Soviet Union reports of brainwashing techniques used on them as captives set the stage.

Over decades, there are many good guys placing their lives on the line, revealing the ongoing high-level official cover-ups of beyond dastardly deeds, such as Dr. Peter Breggin.

To give a better understanding of the mentality of those involved in ongoing human experiments, Dr. Breggin wrote that after World War II, organized psychiatry had begun a program of sterilizing tens of thousands of Americans, and that a patient could not get released unless sterilized in California from a hospital without. And, also, for example, Virginia "Special Needs" children were targeted.

The opportunity to create and change developing minds, then reshaped and molded consciousness as a result is typical of the Nazi styled, fascist and totalitarian agenda.

In fact, Breggin wrote, that a convoy of USA operatives actually went to Berlin from the USA to assist the Nazi sterilization program. This was also while assuring the Nazis that there would be no opposition from the mentally or physically unfit citizens for doing the same to their loved ones by Euthanasia and essentially cold-blooded murder in the United States.

The Association of Psychiatry journal actually published an editorial on Euthanasia, providing guidance during this timeframe stating that psychiatrist would have to "muster their psychological skills" too keep parents from feeling guilty about agreeing to have their children killed.

The two main areas that psychiatrists concentrated on were sterilization and euthanasia. They were responsible for reporting these "patients" over to the authorities and from there, to the gas chambers, killing over 200,000 people deemed mentally ill. These included many thousands of feeble-minded children. The real intention was to rid the delusional and "so-called" master race of "undesirables."

The fact is our journey here on this planet and why we are here is to learn to love. This is although a totally different ongoing agenda is designed to teach hatred.

The murderous effort around me continued and by July of 2018 specifically my lungs, skull, and my breast were now key areas that were being beamed cook.

The lung focus was strategic after a bout with Adult Pneumonia which had been escalated from the Flu in December of 2017. The frequency manipulating technologies have the capability to take a person's body temperature from one extreme to another such as chills to extreme heating and within seconds thus creating Adult Pneumonia.

As a result, of the targeting effort around me, it became necessary to post personal health information on social networks in hope to curtail, by exposure, what the beamed operation was yet again attempting and tests and lab work revealing initial good health or normal healing.

I rationalized that in a program where it is difficult to prove unseen or detectable laser beam weapons used for covert health deterioration, medical documented results and any deviances from typical medical history, as proof of the resulting beamed damage specifically by electronic weapons is vital documentation vital.

My x-ray revealed the right hip surgery on the mend without complication or abnormalities although it was being maliciously recooked to this day, the left as well by the Directed Energy Weapon beam set-up using neighboring locations around me, and it later learned that this surgery, by an experienced VA Orthopedic Surgeon was a botched surgery resulting in a Periprosthetic fracture.

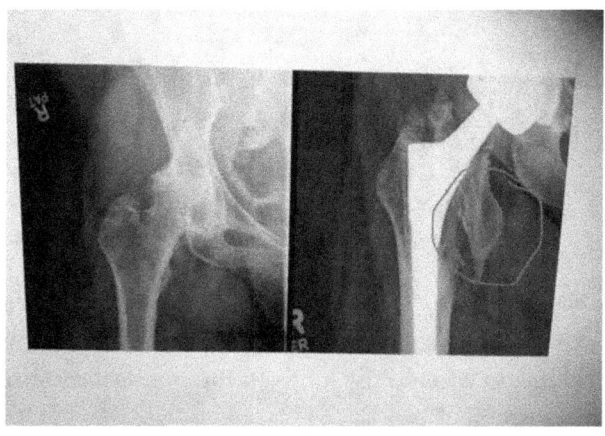

As shown and revealed on x-rays before, left, and on 3/3/2017, right image, there is now painful progressing damage to my right hip prosthetic after apparently forcing the prosthetic into place. Yet the hope was then and today to document no complications within the cover-up by the Department of Veterans Affairs, heavily involved in this high-tech targeting program experimenting on veterans every way possible. In Book III "Covert Technological Murder – Pain Ray Beam" I document how both hip joints were systematically, strategically microwave drone beamed deteriorated with also the portable USAF Directed Energy Weapon portable version, while those at the helm watched as I limped in pain, tracking me 24/7 and using the beamed harassment system, repeating, "Ha, ha, ha… She can't walk."

The documentation of having no complications is a blatant, outright lie! And, some would argue, that the damage was influenced by the VA OIG corrupt team of Special Agents angered at the inability to come into the open unless they can pull off some type of crafty set-up while monitoring my resulting exposure of their bogus operation and covert high-tech human experimentation in this new paradigm today.

> 2/1/18
> Impression:
>
> Bilateral total hip arthroplasty, left side seen on frontal view only and the right one without complication.
>
> A/P:
>   doing very well s/p R THA on 3/3/17. Imaiging shows no abnormalities.

CT scan of the lungs, also heavily targeted after synthetic pneumonia creation was also documented as healing normally. This is although the beam cooked my lungs to the point of creating phlegm in sporadic, intense hits.

> 2/1/18
> Impression:
>
> Bilateral total hip arthroplasty, left side seen on frontal view only and the right one without complication.
>
> A/P:
>   doing very well s/p R THA on 3/3/17. Imaiging shows no abnormalities.

And although new calcification was found on my right breast in the exact same location of the focused drone beam attacks, while I worked on this manuscript, the Mammogram reported there was nothing abnormal as of yet to specific areas of both breasts also targeted.

My next documentation, I realized now required Neuroimaging. Brain beam cooking has also escalated and sometimes to the point of leaving behind deep penetrating pain in my head and an area that is sensitive to touch for days.

A lot of the funding for these operations draws from the secret 52-billion-dollars "Black Budget" and is put to use for silencing efforts by the agencies involved.

This is also the funding source for mobilization or Organized Stalking and paying for use of compliant locations around the target which become effective because of proximity and positioning. Operations around targets can go on for years and until the target's demise.

The effort of strategic silencing and take down continues.

There has also been a specific hope centered around increasing the sleep deprivation effect on me. Sleep deprivation is also massively reported consistently by everyone targeted in this program. For obvious reason, sleep deprivation plays a major role in the hope to break the target.

Sleep is as important to our health as eating, drinking and breathing. It allows our bodies to repair themselves and our brains to consolidate our memories and process information. Poor sleep is linked to physical problems such as a weakened immune system and mental health problems such as anxiety and depression.

The reason intense efforts of sleep deprivation have failed with me is due to medication prescribed by the VA Hospital which insures sleeping through the night. Mediation prescribed for high-blood pressure, of which I do not have, also worked similar to a superior sleeping pill which is why it was prescribed. The end result is a restful undisturbed night.

In my case, the hope for creating sleep disturbance, continued into the wee hour of the morning. The fact is a small degree of my success, aside from sheer willpower, is the result of a balanced restful night. In fact, the only time I knew I was actually under attack, in the wee hours of the morning, as stated Book II, in "Covert Technological Murder – Pain Ray Beam" was only when I woke to use the restroom and felt the irradiation and actually heard the dull hum of the drone Sonic Weapon focused on me and then looked out the window and saw it overhead with lights flashing in the night's sky.

> Impression:
>
> 1. No mammographic evidence of malignancy in either breast.
>
> 2. Further stability of the previously seen benign-appearing asymmetry in the left breast, now considered benign and safe to follow on routine screening.

In my case, I had taken the same medication for over 10 years when in April of 2018, by likely COINTELPRO interference, I instead received Placebo. It is obvious, why I was sent placebo, if you consider that the medication had, as stated above had afforded restful sleep contributing to my sanity and the ability to stand up for myself.

The fact is, if you have taken the same medication for a long period of time, years, and it arrives, a totally different color, taste, texture and smell, it just might be a red flag that something is wrong. To be sure, I tested the mediation for days to determine if it was effective as it typically is hoping that I was wrong or it my imagination.

In fact, a few years prior, I had actually stopped taking it in the morning, because of the instant and extreme drowsy affect it created within 30 minutes which demanded that I get back into bed and nap for at least an hour. When I concluded it was likely placebo, after opening it up and it tasting like powdered sugar, and also after taking it several mornings on an empty stomach, and remained alert, it was clear.

When I told my primary care physician what I suspected, she suggested that I have it tested officially. I did check into this however, due to having to isolate every ingredient of which the medication is made of, extremely expensive. I also rationalized that it would be fruitless to try to get it tested while under 24/7 real-time surveillance and the operation now aware that I knew it was Placebo due to the intense focused real-time monitoring of me without fail.

Great effort is made in this type of covert, deceptive targeting and many strings are being pulled I have learned firsthand from several experiences. Obviously, one had been pulled to have Placebo mailed to my house instead.

Would I be tracked and monitored if I tried, of course I would.

This had happened as tracking me to the Court building in downtown Los Angeles, when trying for legal remedy. I subsequently watched staff clerical staff in the Clerk's office influenced by the LAPD Metro Division, a few blocks away change they responses to me. One of them actually showed up, to set the stage and immediately began reporting to the Clerk's Office that I was in fact seriously mentally ill, and at some level COINTELPRO instruction to hold my submissions to the Court. When the judge stopped receiving my paperwork, it was obvious.

After I found online that there was an actual major placebo study recently testing this specific medication for effectiveness of which I had been sent and it documented as a factual study using placebo by the VA, it was clear. The look of the Pfizer capsules, were identical to the medication shown in the testing. They were a different color capsule, which were pink and white. The ones I received for 10 years from the VA were from Milan Pharmaceuticals, which were tan and brown. This and the guaranteed ineffectiveness were all the proof I needed.

Had I been duped and continued to report medication effective which was nothing but placebo, and took medication of which the collusion sent and reported it effective, it would have been yet another strategically organized set-up which was likely their possible frame of thought. Or, more importantly, if it did not do what it was supposed to do and had done so effectively, I was at high risk for this escalated effort, just simmering hoping for the final shut down, hoping to push me over the edge because of sleep deprivation. Without a doubt, the VA hospital doctor who refilled the mediation, was likely aware of the VA placebo trial, having worked for the VA for 30 years.

He was also aware of the VA veteran targeting program being widely reported, and key also, when I left the appointment that day, after seeing him for the very first time, and several odd remarks he made to me, there sat a familiar face in the lobby. It was the FBI agent around me for some time attempting to hide his face under a baseball cap with his head bowed.

The third incident occurred when I decided that I was not going to see this doctor again, based also on the several red flag comments he made that day, and even before sending the placebo to me as an unwitting test subject. And without my permission which was illegal. I then hoped to see a female doctor as I had always done in the Women's Clinic, although she had left the VA. I had only seen him during a televised appointment as he sat in his office 60 miles away at the West Los Angeles VA facility.

This decision would prove to be a complete disaster for me. It was also how the new clever diagnosis of extreme delusions was officially documented into my medical records one month later. This was by the VA Resident as mentioned earlier. Subsequently, after talking with him, I was floored after the intake and him telling me, after an hour meeting that I was actually in the wrong clinic which logically he should have been able deciphered within minutes.

Again, the only thing reported in his notes which was accurate was my telling him that the FBI and LAPD are targeting me. His notes went downhill from that point on.

Apparently, any type of reporting linked to covert police targeting operations leads to a misdiagnosis of delusions although, a few days later after pulling up his notes online, I was thankful he did put that I was not a danger to self or others. If he had, I likely would have been involuntarily detained and committed on the spot and it all monitored in real-time revealed by numerous horror stories by targets.

Was there a conspiratorial connection? Amazingly, his notes, as a Resident, were in total contradiction to what the 30 staff doctor had reported in his notes earlier and a month prior, although likely involved in switching me to Placebo. He reported that that I was a writer, published several books, etc., well- groomed and well-spoken although telling me his report could change.

Non-bizarre delusions are about situations that could occur in real life, such as being followed, being loved, having an infection, and being deceived by one's spouse.

Across the board when target's report the machination of this high-level government human experimentation program, or being followed, tracked, monitored harassed officially, all targets, with no exception are consistently diagnosed as this being implausible and the target instead severely mentally ill.

Because of the sheer numbers, you must ask why?

And apparently, hoping to stay on top of the game the label of mass delusions has been concocted and is permeating public influence, thus assisting in keeping the truth reported by thousands again, well-hidden and instead highly credible nutcases.

With this happening so much to so many highly credible targets, there is an obvious connection.

The assessment of Delusional Disorder is derived from the implausibility that what a patient is reporting to a clinician is not true or could not even be true. How can a clinician sit in an office, see a person for the first time in their life, nor walk in their shoes, report that factual reality is not?

When a clinician denies the report of true experience, you begin to wonder if something is factually wrong with them and then realize how dangerous they really are. They hold the power of a callously applied mental illness tag, which with some carries the capability for total destruction of lives after the denial of reported factual events. All I could do was leave him with a business card and hoped he looked up my website, etc.

Again, my self-esteem is very high and I don't care what people think. This is especially true when grounded in the truth and dedicated to it no matter the pit falls. Ultimately, the truth will prevail, it always does!

Historically, once a person was labeled mentally ill to include, typically resulting from some type of involvement in a political disturbance or disagreement, the fact is he or she could be sent away to a mental institution indefinitely and, of course, without a trial. And yet again, sadly, the same tactic has reared its ugly head today and is being strategically applied. Can you think of a better way to silence a Constitutional Right to defend oneself after revealing some official injustice, especially after one is now determined instead to be "crazy"?

There was also another incident, alerting me that federal agents were very serious and now going in for the strategic shut down.

In mid-May of 2018, as the threats in the background of my monitored phone, and the drone beamed Hypersonic Sound Device harassment, increased and persisted, I decided after purchasing my home, in 2014 that I had better make legal arrangements for my three daughters, if I were to pass by creation of a Will, or Living Trust and life insurance.

Most agree, the worse thing anyone can do after passing, without making their wishes known, is leave family members, not only in emotional despair, but also in the grief, confusion and frustration. If something were to actually happen to me as the beamed weapon continued its mission, it was now vital.

I had done all I could, to get everything officially documented.

The Pulmonary doctor was informed of what is happening to my lungs trying to heal and being cooked. The Ortho doctors knew from me that my legs and surgical areas of my hips and other joints are being strategically microwave cooked and recooked after surgeries, both in

2012 and 2017. The eye doctor knew that my eyes are sporadically being lightly beamed and my primary care doctor knew of my breast, heart and head were being sporadically cooked by this targeting operation. The only reason people considered me with a degree of credibility was due to the back-up publications.

After inviting an insurance broker into my home, when he left, placed, in a decorative item in the bathroom was a small memory card. I found it because later that night, I felt what I knew to be the directional laser beam of the drone weapon system begin beam cooking my legs coming from under my bed upward, which was over the downstairs bathroom he had used. Knowing it was the likely the drone, I thought I should check anyway, just in case.

A friend has innocently forewarned me not to let just anyone into my house and I thought of him. I wondered that perhaps a small antenna could be used as a directional beam the way a larger one is used to focus the energy weapon, as development of technology continues. I went downstairs just to humor myself and began checking the baseboards, the mirror and medicine cabinet, under the sink, etc.

When I saw the memory card in a decorative item on the counter, at first, I thought it was likely an old memory card possibly holding drafts of older manuscripts for the other books. However, I had always made it a point to keep them, because they are so small and can be lost, together and all in the same place in my home office.

I took it upstairs, and put it into my computer, and was floored to find that on it were numerous images of Islamic Terrorist who appeared to be in some type of Islamic Terrorist training camp. When I saw this, I was mortified and could not believe it such as the images below.

The above images are online images used as an example of near 50 images found on the memory card I found.

Many times, over the years, I have had to check Facebook groups which I had been added to without my permission, who appeared to have Islamic extremist connections or possibly terrorist groups, appearing angry and against the USA, which could have been a masquerade, as well, yet unjoin myself. Facebook allows adding of people to groups without your permission requested first.

One thing I knew for sure, after finding the memory card, I was not going to go to bed with it in my house. I thought about throwing it into the trash, however, there is not much you can do when under 24/7 real-time surveillance and you can hear the lens of the optical technology scratching around in the ceiling and following you from room to room. Instead I decided to torch it. I then created a blog on what happened, and reported it to thousands on social network as a measure of protection.

If there was a raid on my home the next morning, one thing was certain, the only thing THEY would find would be a replica, of a once memory card, holding over fifty images of Islamic people holding weapons, in group settings, which included training of children also holding powerful weapons, now burned to a crisp.

Incredibly, in June of 2018, as I came down the home stretch for publication, I was further shocked when I stumbled upon a Facebook page newly created during this timeframe as well in my name with my Facebook logo and name on it, but in Arabic. This was after I randomly did a Google search for my Facebook page after logging out. Shown below are the results.

Logically, the bogus label of a bogus connection of involvement to terrorist would serve this operation around me well, as surely the mental illness tag would. Both could become highly effective methods, and under the circumstances of what is happening around me expertly for permanent silencing creatively.

How dare I continue to fearlessly fight for my life and ignore the narcissistic ego-driven mania of this program and the operatives high on the perception of their power to destroy anyone to include without accountability and due to a rightful good fight and my courage to tell it all?

The results would be positive only for this operation, by a disk found on my property, and difficulty of me to prove it, backed by the Islamic Facebook page, and no valid explanation of how it got there, would be a brilliant method of entrapment that could break me down they apparently hoped.

Is there a "Silent Holocaust" today?

Renee Pittman Mitchell | Facebook
https://www.facebook.com/reneepittmanm
Renee Pittman Mitchell is on Facebook. Join Facebook to connect with Renee Pittman Mitchell and others you may know. Facebook gives people the power to...
You've visited this page many times. Last visit: 6/12/18

Renee Pittman Mitchell | فيسبوك - Facebook
https://ar-ar.facebook.com/reneepittmanm Translate this page
Renee Pittman Mitchell موجودة على فيسبوك. انضم إلى فيسبوك للتواصل مع Renee Pittman Mitchell وأشخاص آخرين قد تعرفهم. يمنح فيسبوك الناس القدرة على ...

In Natural News, Mike Adams details hundreds of human experiments conducted on the usual victims – the poor, the elderly, mental patients, etc. during the holocaust and over the years since.

If a lot of people are saying the same thing LISTEN!

I.G. Farben, German pharmaceutical company, made use of concentration camp victims to conduct dangerous and fatal drug experiments and were the creators of the deadly gas that killed holocaust victims.

As Target Max H. Williams wrote in a formidable and expertly detailed account of his own targeting in the online article entitled "Silent Massacre / Electronic Stalking and Mind Control in the USA":

"The victims of the Holocaust and today's targeted community provide... excellent examples of individuals who suffered from the torment of this psychological process as, for a variety of reasons, the Nazi's (and FBI's) goal ...is to deprive human qualities such as individuality, compassion and most importantly, identity through the process of dehumanization.

In attempt(s) to create an absolute Nazi hegemony and New Order in Europe (& a new world order forged by the USA) based on the concept of racial hygiene, Hitler (& FBI, CIA, DOD, NSA) and (their) devoted (supporters) began a deadly conquest around Europe (& the globe today) consisting of Hitler's (& FBI, CIA's goal) of seizing and conquering land for the (political realignment/seizures of natural assets) and mass murder and extermination of (non-conforming) people.

With the intention of dehumanizing the (Targets) and naming them as an inferior (FBI, CIA) propaganda successfully created widespread anti- (criminal campaign) which lay down the foundation to eliminate the rights and freedom of the (selected Targets are falsely labeled dangerous criminals who threaten public safety).

The fact is, these types of tactic are nothing new and woven into the fabric of America and have been quite effectively used for many, many years.

CHAPTER TEN

"And in that moment, she realized none of it was real, and so, she set herself free."

~ Katlyn Charlesworth, "Where Men Sit"

Today my heart goes out to the untold numbers of people of all races placed into this program, set-up for many reasons, and uses for total population control stats, jailed, or the human division, by promotion of race wars, programmed hatred, sexualized, and children deflowered and obviously so much more.

I wonder if perhaps, the Frankenstein scientist had stumbled onto a capability to see into the future, and it a proven possibility and resulted in their knowing who would be the problem children, now adults and those who would stand in opposition.

I think about my daughter, whom I love, used against me in the surreal manipulated thought control take down. I remember Myron May, the Assistant District Attorney, Las Cruces, New Mexico, connected to the Florida State University Strozier Library shooting calling me right before going postal at Florida State University saying "He had devised a plan to end this once and for all" and he needed me. This was a plan of which I was never informed.

In gratitude, I am grateful I was unavailable to response to him. This of course would have been yet another perfectly orchestrated set-up if I had with the calls happening right before he went Postal. His eBook is a free download on Smashwords.

I wonder if it was truly his plan or an influenced plan of his Handlers encouraging a young, reported brilliant Assistant District Attorney that going postal was the solution. It defies all logic when you think of it. Going Postal plays into the narrative of mental illness.

Based on ten years of research, it was likely not his plan but, using him against himself self in his thoughts, he may have been led to believe it was. Myron May's plan to end this once and for all, could not stop what has been happening for so long with many lives becoming high-tech sacrificial lambs.

With exposure, it can be done and eventually will be stopped hopefully, but it will take a force, that is greater than the conjured-up evil at play today, backed by the Luciferian globalization agenda in sync and evil puppet-masters and specific agencies.

In spite of their persona, at a covert level, operatives in the program continue to reveal themselves as twisted souls of whom even a defenseless child's life has no value except for scientific research.

In hindsight, many fully aware of mass population control programs as a realistic cover effort operating for decades, and must acknowledge that many human experimentation subjects have been used to bring the agenda, program, and technology to its high state of perfection all over the world.

One thing is certain, I am not Gay and have never had any physical or emotional attachment to another woman in any sexual manner which I find insulting. I love being a woman. Yet, it appeared there was a definite effort trying to create just that with me and it successfully effective with many others. Or, perhaps people are born that way.

As I completed this chapter, the operation who has leased the corner house behind me, using the through the wall radar, for ongoing harassment then combined with the drone focused beam begin issuing new threats of my demise. This was as it began slowly heating my heart from the back and moving to now the left lung, with the right usually the slow cooking preference.

I was so important that "Black Budget" dollars were likely paying for this five- bedroom house of official criminals, and contractors and those I had seen, LAPD, USAF, and COINTELPRO personnel and parolees they are using enter and now were making every attempt to discourage me by the focused beam as a death threat at that moment, and high-tech demand to not publish or else.

Far be it for me to be afraid to fight rightfully for my life!

As stated before, California has been a historic state of mind control testing and the forerunner. In a climate today of nationwide state-of-the-art operation centers, nearly two or in some cases three, both military and federal, state and local police divisions, the "Surveillance State" loomed as the final frontier of the conquering and technocratic control of individuals, groups, communities and large populations.

The JRIC governance definitely, logically, were not pleased with my revealing who is involved in this high-level technological targeting operation, around me today, or my refusal to not be frightened out of fighting back for my life, my children and grandchildren.

No ridiculous mental illness tag was going to stop me from reporting the truth, however if they returned the two cooked away hips joint, and two surgeries, perhaps, I would consider it.

If I suffer a beamed heart attacks, etc., with the beam's focus heavily right now, be it known that this operation is spearheaded from the FBI JRIC, shown in the image above, in Norwalk

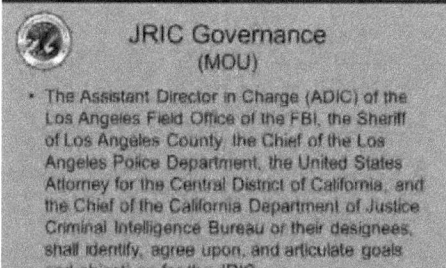

The California approving authority, and Los Angeles County counter-terrorism division who oversees many agencies and who are using military personnel and military technology and who are 100% indirectly to blame.

Them trying to protect exposure of this hideous, inhumane program, with destroying a life, revealed they are more concerned that the truth will hurt images proves who and what they really are as surely as "Covert Technological Murder – Pain Ray Beam" does.

What does "THEY want her" actually means, repeated over and over by the pathetic goons they are using in this game of PsyOps?

It is, a repetitive statement designed to condition their hired killer goons, by approval, for covertly beamed murder and a veiled threat of their lack of employment, apparently if they do not fulfill what THEY want.

One thing is certain, it does not mean, they can come into the open as men of integrity, honorably or respectable. They have had every opportunity in the world to do this. Apparently, THEY choose to hide behind technology instead and attempt to slowly kill, believe it or not, choosing silencing by high-tech murder!

Many have told me that if the government wanted me dead, I would be dead by now. To some degree my proactive efforts have been a life saver, however, the jury is still out on beamed slow kill technology.

My position is clear, in order to fully expose this program, it takes the willingness to place my life on the line. In reality, what choice did I really have? They just don't pack up and leave. Admittedly the wrong being done to me for many years, and also the literal painful reality of truth, connecting the dots, especially after the program was officiated around me and destruction of two hip joint's beam destruction, and ongoing covert health deterioration stood to not frighten me but set my soul on fire.

Hopefully helping others through my providing basic examples of how this program is technologically structured, by furthering understanding of the basic protocol, and the depth of subtle, persuasion mind invasive psychophysical technology, by revealing its use and their techniques, will be able to help some get and grip.

Hopefully awareness will prevent the horrific results which turn victims against themselves, more often than not, traumatically, to include their choosing out, being suicide.

These people are official vampires depleting the victim of their life force during a lifetime and the rest of their lives and their lives now thrown out of sync as expendable human guinea pigs

There is the ever- present hope and pray for release from this hideous new paradigm of which everyone's biometric signature has been downloaded into the system, as reported by the Total Information Awareness Program.

Here are some examples of how it is done.

Have you ever thought about something you never shared with anyone, and believed the thought was between you and God? Have you ever been angry at someone and thought "I could kill him or her" and it only a temporary emotional expression only, knowing you would not do such a thing?

Have ever reminisced about the past, and experiences, which were designed to help you advance and grow emotionally, and shape your perspective positively, and the thoughts are used against you through mind reading software?

Today with the advancement of mind reading technology, a staple in Remote Neural Monitoring programs, and programming, a capability from numerous operation center set-ups across the USA and globally, this is no longer fiction but science nonfiction and the prison system is full of those deemed undesirables.

Technology in research, TESTING, and ongoing develop around the world, for Remote Neural Monitoring has achieved the unimaginable by those at the helm essentially playing God typical of the Devil wanting to be God and working through them.

The fact is it is not the first thought that matters, but the 3rd, 4th, and 5th or more which, as mentioned earlier, draw final conclusions and establish a belief which could also change, with the introduction of a new point of view, new information and experiences.

However, in Remote Neural Monitoring operations, a first thought is being amplified by individuals, who in most cases, have little to no education, military and law enforcement, hired at age 18, working in state-of-the-art operation centers.

Working in shifts around the clock looking for anyway to sabotage the target's life, the key is based on what they are hearing as the target or targets is/are thinking. As I have stated numerous times before, without access to your thoughts, used to gage what is effective, or ongoing hope to locate any weaknesses to exploit or focus on, these efforts would not have the full, traumatic, and devastating impact, on such a personal and intimate level of the target's psyche.

The aspects of mind programming reported nationwide and globally by many reveal's the global paradigm.

- predictive programming definition: inciting an action that causes a predicted reaction

- anchoring definition: the development of mind programming foundations
- conditioning definition: establishing & strengthening neural pathways through repetition
- trigger definition: strong reactionary mind programming related to a particular premise
- association - associative programming (e.g. guilt by association, discrediting by association)
- rationalization - used in psychological warfare to persuade belief & behavioral modification
- belief - beliefs are tied to the mind programming matrix - the mind matrix often serves
- notions that align with one's beliefs. beliefs are a prioritized focus of the black Ops crime syndicate
- projection - people perceive aspects of themselves in others
- perception - Perception is not objective reality - our job is to align the two.
- occulted neuro-linguistic programming (NLP) definition: the mind programming tied to language mind programming details:
- mind programming is universal to consciousness: all living beings have mind programming
- mind programming can be positive or negative (or both)
- every different conceived thought has programming related to it - the natural laws of the mind

- people who have been firmly programmed with deception often have no capability to see reason in a conflicting point of view - character flaws enhance programming susceptibility

Here are a few examples of how Remote Neural Monitoring is being used on Targeted Individuals:

1. You decided to go out to the store or the gym. The operator targeting you listening decides it is the perfect time to beam your legs and cripple you so that you will be unable to walk. The success is in them listening to the target's thoughts about the pain they are inflicting, letting the operation know they are effective in what they are doing and it is working and useful.

2. You decide on eating at your favorite restaurant later. The operator knowing your plans in advance, calls ahead and instructs staff you are on a list as a "Domestic Terrorist, criminal, drug addict, or mentally ill patient and dangerous list, etc., and staff are prepared to play along when you arrive in the Organized Stalking efforts with all monitored in real-time by a drone, satellite guided from an operation center.

3. You are watching television and a love making scene comes on. The remembrance of a normal healthy sex life flutters through your mind before the targeting. The operator then uses the patent that is documented to stimulate you sexually as a form of PsyOps sexual terrorism, and waits for your thoughts on what is happening in the scene to torment you sexually by unwanted stimulation. Or, you look through a fashion magazine and size up other women, something many women do, and think she's looks so nice in that outfit, love the shoe's, etc. The operation then begins sexually stimulating you after misconstruing this thought as being an interest in women sexually and the reportedly the same with any thought about men you are stimulated hoping it will lead to someone they wanted planted around you.

4. You don't like of bugs in your home, make extra efforts towards cleanliness, and find them disgusting in thought. The operator then decides that one way to bring you down, and traumatize you would be to release a bug infestation inside your garage, then beam bug nightmares.

5. You develop a fear of heights you never knew you had before. This is downloaded into the supercomputer perhaps one day when you are on a ride at an amusement park, when the feeling is registered. When an opportune time arises in the future, and you are in a high place, those monitoring you, who have also cloned and uploaded many and various emotional responses through the monitoring, then beam the computer cloned emotions back as an electromagnetic frequency and you become excessively fearful.

6. You think about your children and decide to call them later. The operation targeting you for years, typically want to isolate ALL targets and keep them that way, and have also downloaded your family's biometric signature, and that of everyone within the US, done by the Total Information Awareness Program, immediately access the family member's biometric signature, connect to them wherever they are on the face of the Earth and begin RNM on them by implanting negative thoughts about you or even anger thoughts about you even before you even pick up the phone.

7. While out in public, the thoughts of people around you are being read, as the real-time focus follows you from location to location, and people are being beamed that they should not like you, or various other nonsense, for whatever reason. People start acting in a manner in which is rude, out of the blue, and they have not a clue of how are why and you do not know them and never met them before.

8. You think I hope they don't give me breast cancer. The operation then believing it is a useful thought to frighten and

horrify you, then begins focusing the microwave Directed Energy Weapon on your breast while still monitoring you for fear and a reaction that can be amplified with repeated attacks and used to break you down and submit. Or the beam is repeatedly focused on the inner thigh at the same location for months, with your hip joint slowly deteriorating and the microwave effect depleting fluids. This is also a sign to these operations of success that then escalate the torture to specific areas because you are concerned and can now be possibly controlled by the thought of death or destruction.

9. You come home late one night and enter a pitch-black house; a thought crosses your mind that someone might be inside waiting to murder you beamed. The satellite then begins to make sounds as if someone is opening a door or walking around upstairs.

10. You are working, and deeply focused on a project, the doorbell rings giving you a startle reflex. They register fear due to you being startled and heart briefly pounding. You are beamed via RNM that it is the police finally there, after continual threatening. The satellite beam then beams a door bell sound but no one is there another day just to see if they have finally found something. Fear is a powerful control mechanism.

11. You are relaxing at home, once a private sanctuary, comfortable, t-shirt and sweats, your hair frazzled, and not looking like a beauty queen but loving yourself and comfortable as the real you. You look in the mirror and the operator / buffoon remotely monitoring you then says, literally "She ain't that cute" beamed from a location in the ceiling. You think? Why does this idiot care? He cares because he is an idiot, this is his version of so-called brilliant PsyOps although too stupid to realized, his childish are foolish and that he is on the wrong team of evil intent.

12. In some cases, some of these operators will simply sit for his entire eight-hour shift repeating back or mimicking everything you think and believe they are intellectual giants by doing so.

13. One day you and one of your daughters were fussing about something that was nothing serious at all, however, she becomes more agitated than you. You distinctly hear one of the individuals watching everything in real-time, sounding as if projected in mid-air, try to entice a mother daughter fight by saying "Beat her up!!!"

14. Your hands are being beamed cooked believing it will stop you from activism, exposure, and typing and creating information etc., or material on your computer.

15. If these operations register that you like your hair, or people notice it and think the same, they will beam your head hoping to burn your hair from your scalp. While it is growing back, every you go they beam, "That's not her hair," clowning but seriously hoping to change how people look at you if they register people find you attractive by acting like jealous females themselves although supposedly men.

16. These operations will attempt to have keys made to your home Again, who are they really protecting and serving?

Again, to Protect and to Serve Whom?

As stated I chose to write books as my weapon of choice and the operatives have continually, over the years requested that if I stopped this book series they would leave me alone. I highly doubt this to be true and law enforcement can tell you anything, legally to achieve results.

I can guarantee that if I reacted negatively in any manner or became confrontational, I would likely be shot down in the middle of the streets, and a brief news report would report that Renee Pittman had been diagnosed as delusional and was a danger to self or others such as what has been publicized regarding all others pushed over the edge and to the brink of destruction.

I had to ask myself did I ever really have a chance in life. I definitely like so many had been manipulated and influenced throughout my life by the subtle influence of the Sun and wind by whomever was spearheading the mind control program in Los Angeles beginning in the 60s and in the early stages.

One day in elementary school, I was called from the classroom and taken to the principal's office around age six. In the room was a psychologist from an outside agency. He began giving me a series of test of which I zapped through quickly. He wanted me to prove I could do it. However, when I did the look on his face appeared to register that he was not happy at all with my doing so well and also confusion.

The shoulder patch of the LAPD Metropolitan Division.

Common name Metro Division

Abbreviation LAPD

Motto "To Protect and to Serve"

This was during a time when there was an ongoing debate of whether whites were more intelligent than blacks and that blacks only had muscular strength and are brainless.

Even today in this program, the men around me, have repeatedly said, "They" meaning their supervisory staff, "can't believe she is black" which is utterly ridiculous when compared to the I know of all races. One even said, "She is making us all look good" as he turned the microwave beam to a higher level intensifying the pain in my legs, or the operation alternates with USAF personnel, or the FBI shows up dispatched from, it appears, dispatched from the residential satellite office about 15 miles away in a nearby city.

The result of the targeting over the years was my barely able to walk before surgeries and excruciating pain as the effort cooked my hip joints. While inflicting this, as stated the military continues to play a pivotal role and even by showing up for weekend duty, nearby, to include federal agents using strategic locations and appearing to actually want me to see them.

The fact is, beginning from childhood, and by subtle influence, there are many types of programming, and even children can be groomed to be racist, and mass shooters, and more as putty in the hands of mad scientist who in actuality make Dr. Frankenstein's work look like child's play. People are taught to hate.

CHAPTER ELEVEN

Sheeple vs. Sheeple - Crafted by Wolves

The servants of the global ideations are factually sitting in senior positions and operating at all levels of government. This is said to be from the highest levels, arguably to the Commander-in-Chief, Congress, the Senate, judges to directors of the NSA, CIA, DOD, DOJ, FBI, DHS and senior leaders of law enforcement and all part of the brotherhood and paying homage by what is viewed as a secret government operating behind the scenes.

The people whose minds have been controlled are reportedly used by the military as sex slaves. They can be used to blackmail politicians. They can be used to infiltrate organizations. They can be used to carry out assassinations or suicide bombings.

Reportedly, when the military are looking for people whose minds they can control, they look for people connected to orphanages, foster care homes, juvenile facilities, families linked to military intelligence, families with adopted children in my case, families interested in Satanism and families involved in child abuse.

Reportedly, many of the mind-control victims come from families linked to certain forms of Catholicism, Mormonism, or charismatic Christianity.

The public has been bought and sold on the need for counter-terrorism divisions, which are factually being used to stifle any protest of what is really happening today based the historic capability to mobilize mass consciousness and hopeful reform against the goals of an obvious evil agenda by Globalist and psychopathic global puppeteers.

Because of the unspeakable cruelty of this program, many in disbelief believe that those enforcing what is happening must be non-human and I have heard many question if they are factually human at all, but instead disembodied evil spirits who have taken over their minds and bodies. However, many agree that what happening today is definitely satanic. However, it is not Aliens or demons as physical entities, at the helm of the technology, but those possessed and programmed by spiritual wickedness.

The hope is the manipulation of the electromagnetic energy field to keep humanity in a lowered frequency and in this lowered dimensional state of being programmable. At this frequency, death, murder, war, and other horrors and atrocities are acceptable arising from lowered consciousness where human compassion, empathy and love has been strategically annihilated by intense highly effective propaganda and repetitive endless programming techniques creating our reality.

Be aware, unmatched and equal to none in the mass population control dynamic is our favorite pastime, the television and news programs.

In recent studies television propaganda is confirmed as assisting in creating a mind-controlled culture backed by military silent weapons, frequency influence, and together a highly effective combination.

Walter Lippmann, an American intellectual, writer, and two-time Pulitzer Prize winner brought forth one of the first works concerning the usage of mass media in America.

In "Public Opinion (1922)," Lippmann compared the masses to a "great beast" and a "bewildered herd" that needed to be guided by a governing class. He described the ruling elite as "a specialized class whose interests reach beyond the locality." This class is composed of experts, specialists and bureaucrats. According to Lippmann, the experts, who often are referred to as "elites," are to be a machinery of knowledge that circumvents the primary defect of democracy, the impossible ideal of the "omnicompetent citizen." The trampling and roaring "bewildered herd" has its function: to be "the interested spectators of action," i.e. not participants. Participation is the duty of "the responsible man", which is not the regular citizen.

He further stated that mass media and propaganda are therefore tools that must be used by the elite to rule the public without physical coercion. One important concept presented by Lippmann is the "manufacture of consent," which is, in short, the manipulation of public opinion to accept the elite's agenda. It is Lippmann's opinion that the general public is not qualified to reason and to decide on important issues. It is therefore important for the elite to decide "for its own good" and then sell those decisions to the masses.

The television and electromagnetic neuro-weapons today are two instruments capable of global hypnotic perceptions and creation of a changed version of reality, backed up and bombarded by subliminal imagery dispensed through television broadcasts that effectively penetrates millions of American households around the clock.

The goal of media, is to intentionally shape, mold, and hone our ideas, attitudes, thoughts, opinions, and insure that, "We the Sheeple," are compatible with the New World Order goals and insure that our thinking is in sync if not you can become targeted.

It is no secret that the most expansive use of media today is focused on propaganda, beheading, terrorist, for example, promulgated, which typically deal with mobilization of war and not peace. Propaganda is influential because everyone is susceptible to it and can be influence by it. The images of the media become a hypnotic imprint within our consciousness.

When official "MKULTRA" mind control studies, documented to be from the 1950s to 1970s came light and then under investigation by the Senate Church Committee Hearings many documents were ordered destroyed. However, many survived and were accessed via the Freedom of Information Act and publicized which snowballed awareness.

Today we are living, it appears, the results of some of the seedier efforts, of the 149 Subprojects of MKULTRA of which then CIA Director Richard Helms ordered destroyed.

If the focus of this specific timeframe, which set the stage, in the history of humankind, is accepted as the only reality, our view of the world is actually limited within the Universal experience.

As the great shift of consciousness is upon us, some believe that thousands of incarnated souls are here for this special occasion and with a specific purpose after signing up and accepting various challenges refusing to be unclothed by the Sun or the Wind. What a nice thought. I have spoken with many who are primary targets of this program, and concluded that some victims are targeted because of some type of spiritual gift or pureness of heart found objectionable as the Devil's influence walks to and fro.

Whether fact or fiction, rest assured that Divine Order has calculated humanity leaving the Third Dimension, crossing briefly through the darkness of the Fourth Dimension, then entering the Fifth Dimension on course. Earlier songs of were written about this order and shift such as "This is the Dawning of the Age of Aquarius" by a group name the "Fifth Dimension." Or some call it the "New Earth" biblically.

The Fifth Dimension is not a New Age concept or tied to the NWO as a secret society doctrine which continues to deceive humanity by promotion that their purpose is in congruence with Divine Order and brotherly love, and godly intent or part of the human evolutionary process ordained naturally by God. Our heavenly blueprint demands that we evolve. This program, up a ladder to the highest level is hoping to stop the consciousness advancement, the Awakening.

In the excerpt below, J. O'Brian writes from his website O'Brian's Extraterrestrials:

The Fourth Dimension - The Truth...The Astral Plane

"The Fourth Dimension is a gray, polarized plane, housing the forces of Light and Darkness.

The battle between good and evil starts here. Forms naturally morph on the Astral Plane... a tree can easily transform into a wolf. This is because the illusion of good and evil is manifest here and because of the extreme mutability of form, distrust and fear that exist... e-motion.

This plane is that of Will or Life-spirit and it is of this dimension that the individual "self," the Ego, is a part. It is the Ego who uses the physical, astral and mind bodies as tools with which to achieve its purpose. When mind, body and spirit are completely aligned with Divine Will and in harmony and balance... one with another... you are omnipotent and have achieved conquest over matter.

After careful training, it is possible to leave the physical body safely in the nourishing care of its etheric web, to go on a Dimensional Journey. This is called astral travel.

When you wish to return you slip back into the restricting burden of your outer coating of flesh. Shamans are adepts in this arena, many times bringing back information to benefit humankind. Unless the person has been specially trained and practiced, the jar of contact once again with the dense Earth vibrations are so harsh that it usually snaps the thread of memory of the journey. Magic, time travel, karma, reincarnation, luck, psychic surgery, flying, mind reading, disembodied spirits, enchantment, and of course, astral travel, all source from this plane.

The Demi-God/desses of many religions live here. Hell, and purgatory are fourth dimensional locales as well. By embodying the principles of this plane, we enhance the probability of living a magical life..."

The fact is both good and evil are energy, or negative and positive. The fact is we can choose which energy we want to resonate too. Yes, it will appear to get worse before it gets better, but better it will get.

I agree, on a super-conscious level, and especially good and evil as energy: Ephesians 6:12

"For we wrestle not against flesh and blood, but against principalities, against powers, against the rulers of the darkness of this world, against spiritual wickedness in high places."

The key to the entry of good and evil energy is the use of the electromagnetic spectrum, it nothing new and its use dating back to ancient times.

As we embark on the Age of Enlightenment, and enter the Fifth Dimension, leaving the Third, the Fourth becomes merely a stepping stone of which only those aligned and resonating to the highest frequency, love, kindness, compassions, brotherly love etc., fighting for good and to make this planet and the world a better place will arrive relatively easy during the Global Paradigm Shift.

Hang in there! Giving in to efforts to lower your frequency aligns you with the goal of collecting souls.

Live the love frequency. There is no greater frequency on this physical, material plane, or prevalent throughout the entire universe. Evil has run its course, and those in allegiance to it, know their time is running out thus are wreaking global havoc. Remain faithful. Soon they will regrettably understand that the fruitless escalations of the Luciferian agenda are just that fruitless.

Real power is in attunement with the higher energy of Light and not darkness. The high-tech explosion of highly advanced technology again, is their hope to assist and stop what is ordained. For them, it will become null and void in the hope of maintaining the trance like state over humanity.

I and millions of others prove the opposition or adversary is fruitless as the veil continues to lift. And it appears; those that refuse to bend and align themselves with the evil or have been able to break the spell have become the threat and are the focus, in some cases, of a targeting program so monstrous, most cannot believe it exists as a global dynamic.

The "Five Eyes", often abbreviated as "FVEY", refer to an intelligence alliance comprising Australia, Canada, New Zealand, the United Kingdom and the United States. These countries are bound by the multilateral UK/USA Agreement, a treaty for joint cooperation in signals intelligence.

What are the Five Eyes aka Satanic Eyes?

The Five Eyes alliance is a secretive, global surveillance arrangement of States comprised of the United States National Security Agency (NSA), the United Kingdom's Government Communications Headquarters (GCHQ), Canada's Communications Security Establishment Canada (CSEC), the Australian Signals Directorate (ASD), and New Zealand's Government Communications Security Bureau (GCSB).

Beginning in 1946, an alliance of five English-speaking countries (the US, the UK, Australia, Canada and New Zealand) developed a series of bilateral agreements over more than a decade that became known as the UKUSA agreement, establishing the Five Eyes alliance for the purpose of sharing intelligence, primarily signals intelligence (SIGINT). For almost 70 years, this secret post-war alliance of five English-speaking countries has been building a global surveillance infrastructure to "master the internet" and spy on the world's communications.

What does the Five Eyes agreement say?

Despite being nearly 70 years old, very little is known about the alliance and the agreements that bind them. While the existence of the agreement has been noted in history books and references are often made to it as part of reporting on the intelligence agencies, there is little knowledge or understanding outside the services themselves of exactly what the arrangement comprises.

Even within the governments of the respective countries, which the intelligence agencies are meant to serve, there has historically been little appreciation for the extent of the arrangement. In fact, it is so secretive that the Australian prime minister reportedly wasn't informed of its existence until 1973 and no government officially acknowledged the arrangement by name until 1999. Few documents have been released detailing the Five Eyes surveillance arrangement. To read the documents available, click here for the National Archives and here for the NSA's release of the UK/USA Agreement.

Here's what we do know: under the agreement interception, collection, acquisition, analysis, and decryption is conducted by each of the State parties in their respective parts of the globe, and all intelligence information is shared by default. The agreement is wide in scope and establishes jointly-run operations centers where operatives from multiple intelligence agencies of the Five Eyes States work alongside each other.

Further, tasks are divided between SIGINT agencies, ensuring that the Five Eyes alliance is far more than a set of principles of collaboration. The level of cooperation under the agreement is so complete that the national product is often indistinguishable.

What's the extent of Five Eyes collaboration?

Together the Five Eyes collaborated and developed specific technical programs of collection and analysis. One senior member of Britain's intelligence community said "When you get a GCHQ pass it gives you access to the NSA too. You can walk into the NSA and find GCHQ staff holding senior management positions, and vice versa. When the NSA has a piece of intelligence, it will very often ask GCHQ for a second opinion. There have been ups and downs over the years, of course. But in general, the NSA and GCHQ are extremely close allies. They rely on each other."

The close relationship between the five States is also evidenced by documents recently released by Edward Snowden.

Almost all of the documents include the classification "Top Secret//COMINT//Relations to USA, AUS, CAN, GBR, NZL" or "Top Secret//COMINT//Relations to USA, FVEY." These classification markings indicate the material is top-secret communications intelligence (aka SIGINT) material that can be released to the US, Australia, Canada, United Kingdom and New Zealand. The purpose of the relationship is to identify classified information that a party has predetermined to be releasable (or has already been released) through established foreign disclosure procedures and channels, to a foreign country or international organization.

The level of co-operation under the UKUSA agreement is so complete that "the national product is often indistinguishable." Another former British spy has said that "cooperation between the two countries, particularly, in SIGINT, is so close that it becomes very difficult to know who is doing what [...] it's just organizational mess."

Despite rumors of a "no-spy pact", there is no prohibition on intelligence-gathering by Five Eyes States on the citizens or residents of other Five Eyes States, although there is a general understanding that citizens will not be directly targeted and where communications are incidentally intercepted there will be an effort to minimize the use and analysis of such communications by the intercepting State.

Are there any other surveillance alliances? In addition to the Five Eyes alliance, a number of other surveillance partnerships exist: 9 Eyes: The Five Eyes, with the addition of Denmark, France, the Netherlands and Norway; 14 Eyes: the 9 Eyes, with the addition of Germany, Belgium, Italy, Spain and Sweden; 41 Eyes: all of the above, with the addition of the allied coalition in Afghanistan;

Tier B countries with which the Five Eyes have "focused cooperation" on computer network exploitation, including Austria, Belgium, Czech Republic, Denmark, Germany, Greece, Hungry, Iceland, Italy, Japan, Luxembourg, Netherland, Norway, Poland, Portugal, South Korea, Spain, Sweden, Switzerland and Turkey;

Club of Berne: 17 members including primarily European States; the US is not a member;

The Counterterrorist Group: a wider membership than the 17 European States that make up the Club of Berne, and includes the US;

NATO Special Committee: made up of the heads of the security services of NATO member countries.

"The Five Eyes Alliance" between Britain, USA, Canada, New Zealand, and Australia?

Not only are people speaking out within the United States about Mass Control Technology, but Canada, Brazil, Japan, UK, Germany, South Korea, New Zealand, and as shown Sweden.

Worldwide thousands of civilian targeted individuals suffer from the illegal experimentation.

Since the discoveries of Nikola Tesla in radar technology and their world- wide spreading, especially first world countries enhanced their own programs to influence electromagnetic fields in the environment.

During World War II German Scientists improved the technologies in applied research using Concentration Camp prisoners for inhuman experimentation.

Nowadays the technological development has risen to unexpected heights.

The National Security Agency (NSA) possesses Nano computer technology that is supposed to be 15 years ahead of the world's development.

These entities can affect any electromagnetic system on the globe such as bioelectrical fields in the human body, any computer or telecommunication device either geological systems.

The effects of this intrusion are serious illnesses, global surveillance, altering human brainwave (mind control) and environmental catastrophes including earthquakes, storms, flood (tsunamis) and volcanic activity.

Worldwide thousands of civil targeted individuals suffer from the illegal experimentation.

Great Britain, India, Switzerland, Ethiopia, Germany, Spain, Netherlands, Trinidad and Tobago, Thailand, Finland, Croatia, Denmark.

"A nation can survive its fools, and even the ambitious. But it cannot survive treason from within.

An enemy at the gates is less formidable, for he is known and carries his banner openly. But the traitor moves amongst those within the gate freely, his sly whispers rustling through all the alleys, heard in the very halls of government itself.

For the traitor appears not a traitor; he speaks in accents familiar to his victims, and he wears their face and their arguments, he appeals to the baseness that lies deep in the hearts of all men. He rots the soul of a nation, he works secretly and unknown in the night to undermine the pillars of the city, he infects the body politic so that it can no longer resist. A murderer is less to fear.

"The traitor is the plague." -- Marcus Tullius Cicero

Henry Kissinger: "Military men are just dumb, stupid animals to be used as pawns in foreign policy."

David Rockefeller: "Some even believe that we (the Rockefeller family) are part of a secret cabal working against the best interest of the United States characterizing my family and me as "internationalist" and of conspiring with others around the world to build a more integrated global political and economic structure – one world if you will. If that's the charge, I stand guilty and am proud of it."

"We are on the verge of a global transformation. All we need is the right major crisis and the nations will accept the New World Order."

The New World Order did not die with Rockefeller...

The question is: "Is strategic, manipulated and influenced "civil unrest" being cleverly orchestrated, right before our very eyes, to usher in yet another official stage of control such as Marshal Law?"

Time will tell.

One thing is certain, if there is civil unrest, it could be the result of intentional beamed agitation focus on populations designed to result in official approval for the next phase.

A lot of the war, death and destruction on the Earth are motivated by and because of a specific type of men and mentality or "Order Followers" carrying out the agenda of Order out of Chaos.

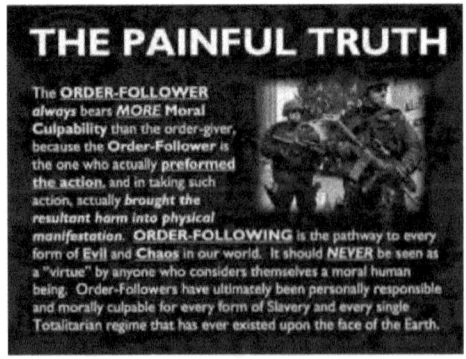

They appear easily programmable into a whole different reality and stuck there. As long as their actions are guided by others giving them orders of destruction, they believe themselves unaccountable.

That is a rhyme and reason that this evil is after females. It knows that as vehicles for their birth, through intense pain, women are capable of healing men from the evil influence and can change them back to their real selves.

Is really the adversary influence of the Global Elite / Puppet Master is against ALL of us?

We must understand the Universal technological aspect and the capability of the technology weaponized in the truest sense revealed by Tesla's comment,

"If you want to find the secret of the Universe, think in terms of energy and frequency vibration."

Tesla also said,

"It is paradoxical, yet true, to say, that the more we know, the more ignorant we become in the absolute sense, for it is only through enlightenment that we become conscious of our limitations. Precisely, one of the most gratifying results of intellect evolution is the continuous opening up of new and greater prospects."

However, I am sure Tesla, a reported Humanitarian by nature, would not agree with the technological use and abuse of patented technology derived from honing the energy of the Electromagnetic Spectrum today and likely would turn over in his grave. Knowledge should not be hoarded for useful control but used for the advancement and betterment of humanity.

Their goal continues in the search to maintain control.

The nature of the research included focus on what was determined these specific problems:

- Can we create by post-H (hypnotic) control an action contrary to an individual's basic moral principles?

- Can we "alter" a person's personality? Can we guarantee total amnesia under any and all conditions?

- Could we seize a subject and in the space of an hour by post-Hypnotic control him crash an airplane?

Can we devise a system for making unwilling subjects into willing agents and then transfer that control to untrained agency agents in the field by use of codes or identifying signs? NB 13, 14, 28 [5]

In a 1971 Science Digest article [6], Dr. G.H. Estabrooks states, "By the 1920's clinical hypnotists learned to split certain individuals into multiple personalities like Jeckyl-Hydes.

With mind control testing dating back centuries, it is clear, we have always been considered, at some level, little more than Sheeple.

CHAPTER TWELVE

Vibrate at the Highest Frequency - LOVE

One of the major elements in modern brainwashing, which has been around since WWII, and perhaps longer, in its modern form, is that of fairy tales, superhero or vigilante.

To young children, this is something to aspire as they sit before mesmerizing television on Saturday mornings watching the cartoon channel. To them, the implication is that this is the pinnacle of being good. When the law is blurred, with Civil and Constitutional Rights denied by those viewed as superheroes, this reflects the deterioration of the moral code and results in creation of an unjust society.

Today the twisted image of heroes is specifically applied to those approved to use the technology that unquestionably has been transformed into something beyond belief inhumane and hidden. Yet, our perceptions have been implanted so deeply of them as heroes, that their denial of rightful, humane, treatment of others is a blurred line of acceptance.

In essence the superhero and vigilante can be viewed in the dynamic of this covert targeting program as an anti-American vision which goes against what basic Human Rights stands for and hopes to insure by fair treatment of human beings.

Unlike vigilantes and superheroes of movies, when humans seek this type of power, it can be an act of selfishness and hope to use this perception and attach the image to oneself. The determining factor, of good or evil is the hideous dissolution of very basic of rights, and more horrifically on a secret, well-hidden level that today has become something entirely opposite of honorable. The long-term consequences, unfolding right before our very eyes, are a natural decline in social standards and moral decay.

This program through years of strategic efforts has today evolved into an accepted reality which allows "so-called" groups of people labeled as undesirables to be mass murdered in slow kill operations in a flashback to Nazi eugenic programs as history repeats itself. Years of programming has effectively created, some, not all in our society who lack the very thing that makes us human, our humanity.

In this respect, vigilantes and super heroes are really proto-fascist characters of unrealistic characteristic types that go beyond any special ability they are purported to have to save humanity, in this case advanced technology, but instead ego-driven entities focused on destroying.

In George Orwell's "1984" a mirror of what is happening today is revealed when he writes about those we are programmed to believe are super-heroes who really are not nor do they have our best interest at heart. Their allegiance is to different belief and concept.

George Orwell writes in "1984":

> "Members of the Brotherhood are prepared to: give one's life; commit murder; commit acts of sabotage; betray one's country to foreign powers; cheat, forge, and blackmail; corrupt the mind of children; distribute habit -forming drugs; encourage prostitution; disseminate venereal diseases; do anything which is likely to cause demoralization of society."

When we scale this to an entire population, that mindset is one that accepts torture, kidnapping, assassination, destruction of the soul of children, or a global event to prop up their economy with the benefits for only and redirected to Globalist at the cost of lives.

This, obviously, is not done as a single step. It is something that gradually seeps in over time, or for example by a catalyst, like the 9/11 attacks, which can then be leveraged to bring those ideas to the forefront by pumping up the programmed population and crowd with calls for revenge and aggressive action deceptively promoted as rightful while playing on the emotional charge.

For those blessed to uncover his or her role today targeted, and awoken, experts believe that there are many more who will never know that their lives have been part of a synthetic Matrix, Hell bent on covert destruction of free thought by influence, suggestions, and self-sabotage programming to act out self-destructively and negativity. All lives matter!

This includes those brought down to nothing, destroyed and homeless, or strategically housed in psych wards and falsely imprisoned. It would seem that testing would be finalized, after the destruction which nonconsensual human experimentation sought to achieve. However, this is not the case, in actuality testing continues. The program knows that most started with a sound mind before the outside source influence and that it might return to self, or what is left of self, if left alone and possibly become an enemy the covert objective and heinous injustice.

In these opposing signs and times of horrifically wrong destruction, a clear invitation lies ahead for us to see who we are, individually and as a whole, and how our past definitions of ourselves, and our intensely programming definition of ourselves, both overtly and covertly has been radically altered or created.

Reportedly, a Bishop in America went viral, when he claimed that Hell is a creation of church and religion to straighten the errant that would, otherwise, be not be afraid to go astray from the laid down path. Understandably, the devilish people, having reached the Kingdom Come have only Hell to give them the refuge.

Sadly, it has been tried, tested and proven that a certain type of man needs some type of moral compass, which has been proven to keep him humane by fear the fear of Hell as a specific location of brimstone that served this purpose, however some would argue today, this threat is no longer effective in the awakening of both good and evil.

If one is in a tight corner, surrounded by evil doers, thugs and self-seeking usurpers, the conclusion may be automatic, "The Devils are here," and "from where would they come?" The answer: "Could it be from their abode, and the dark energy of Hell infused in their consciousness and guiding them?" As a logical corollary, then Hell would wear a deserted look because as Shakespeare reported, "Hell is empty and the devils are here." And their sole / soul's purpose is to remain on the hunt for any energy different from their own because they first seek, a narcissist validation, of the righteous of evil they emit and of which they perceive as normal, conceived in their hearts and minds.

There is a somewhat similar idiom: 'To let the Hell loose.' The meaning and affect are almost similar. The heinous creation of destruction of innocence, beauty, that which is good and pure, torture, inflicted depression, pain, suffering and sorrow results in suicide, which is their empowerment.

If you give a man the correct information for seven years, he may believe the incorrect information on the first day of the eighth year when it is necessary, from your point of view, that he should do so.

The first job of the program is to build the credibility and the authenticity of the promoted propaganda and persuade the enemy, you, to trust it although you are unknowingly unaware that you are factually the enemy.

Above all other strengths and the ultimate success with me has been, as with many others, the determination to always self-correct, and search for greater understanding and to not let this program make me bitter and hateful. I cannot operate positively on a lowered frequency. And if so, I would begin to operate on the same level of the enforcers.

It is easy to become them or become like them through their constant heinous PsyOps and psychophysical torture, constant denigration and high-tech influence around targets. It is a constant battle to fight for good and positive spirit.

This does not mean I won't forget this cruelty or the reality of what I am up against! Instead, personally I will continue to transmute the darkness of those around me and those captured by the darkness into light for my personal self-survival by thought, word and deed. I cannot allow myself, because it is not my true nature to be changed negatively. This is because of my awareness that thoughts manifest reality and thoughts are, and become, things.

The hope is to align purpose by the will of God first and above all else of which the "Compendium of Wisdom" explains and describes.

Knowledge is possessed only by sharing: it's safeguarded by wisdom and socialized by love. The mortal soul does not survive by what it does, but by what it strives to do.

Difficulties may challenge mediocrity and defeat the fearful, but on stimulate the true children of God.

The greatest affliction in the universe is to never have been afflicted. Mortal man only learns wisdom by experiencing tribulation.

Effort does not always produce joy, but there is no joy without effort.

The weak indulge in resolution, but the strong act. The act is ours; the consequence is G-ds.

To enjoy privilege without abuse, to have liberty without license, to possess power and use it for the good of others~ are signs of high civilization.

Affliction is the ridiculous effort of the ignorant to appear wise, the attempt of the Barron soul to appear rich.

Anxiety must be shunned, the disappointments hardest to bear are those which never come. I'm patience is a spiritual poison: anger is like a stone thrown into a hornet's nest.

The augmentative defense of any position is inversely proportional to the truth contained.

The destiny of eternity is determined moment by moment by the achievements of day by day living. The act of today is the destiny of tomorrow.

Blind and unforeseen accidents do not occur in the universe. Neither do angels assist the mortal man that refuses to act upon the light of truth.

The archangel of the universe

I believe it imperative to maintain balance.

We can and will beat this program and change the world through the darkest of challenges. These thoughts are what I hold onto with a surety.

Targets that are courageously awake today don't suspect foul play in our lives but from extreme experiences are, prepared for it, know it is very real, technologically, and cope around the clock. Sure, the program, is logically difficult to believe, however take the word of a unified effort with the United States and globally near millions who are reporting, this program is very real and has been for decades.

I learned to flip the switch, on those involved after what appears to be their very own specific type to a group "Herd Mentality" or "Hive Mind" programming, who appear oblivious that they are also likely being subtly being manipulated and influenced into committing atrocities. Use of them in this program demands it.

It is Aesop's Fable played out by mind invasive, technological challenges focused on humanity scientifically as the battle of strong wind challenges or the Sun's appearing subtly and harmless influence. The fact is both were control orientated and the battle is for our minds.

Today's generation is the product of expert desensitization to violence over many years by television for one and violent video games. However, many people are connecting the dots and without a doubt anyone by the use and power of positive energy can become dangerous, to the program, and to the heinous grand scheme.

We are all being used in one way or another and one thing is certain, and many can guarantee, we are being watched in a massive Orwellian "Surveillance State" with access to highly advanced officially patented advanced technology.

During another incident while running errands, the beamed influence of a person got so bad around me that I watched the manager of a store struggling with intensified anxiety manifested, and my knowing what was possibly happening. After observing her strife for a year or so, I told her of the likelihood of beamed tracking around me and the influence of those targeting me. Because I am never one to believe my assumptions are always right, I could be wrong. I have asked her if she was having negative thoughts specifically around me. Her response was an immediate and yes, and that she also is very familiar with negative thoughts.

I observed her trying to block the possible beamed influence by humming as she worked or begin to sing Gospel songs, which started each and every time when I showed up.

As a Christian, she admitted the negativity, and was relieve to know that there was a, possible outside source, at the helm of the bombardment of which I had become familiar with while being tracked around the clock from a state-of-the-art operation center, and by use of one of 30,000 drones approved for US skies by US Congress by 2020.

Imagine today millions being influence to some degree or another this way. When you do, it is heartbreaking, and you understand that people have a right to know, no matter what the consequences, about the charade of this massive program to give them a fighting chance.

It was only because of the desperation to silence me, and overall arrogance of this program believing myself and again, millions of people would not figure it out. And more importantly since targeted as a child this effort was also revealed as a totally bogus official investigation around me. THEY have wasted a lot of time, energy, and resources here.

With me this operation cannot back down especially when I have made it my mission to expose this program. The only choice is for them to live the lie, convince themselves they are good or, heroes, so that they can do what THEY have ordered to do.

I thank God for the same strong will, and spirit, this operation has been unable to influence or break even as child and throughout the years which ultimately enabled me to complete this book series under the most treacherous of circumstances.

THEY want the satisfaction of the subject jumping through hoops, cowering, frightened, immobile, and scared to live. You become and your life becomes a toy for conceit and vicious arrogance.

Do not put your faith and trust in the programmed order followers used in this program. They can do absolutely nothing with the truth. The truth would short circuit their programming of which has essentially relegated them to red blooded Westworld robots.

There is nothing they can do with the truth after hired for these positions, and job description of employment clear. The only thing they can do is do what their Handlers request, or in their case programmed into their psyche, to allow them to torture and murder for them. And while doing so earn a living then go home to their families while destroying others secretively.

In reality there is a connection to misguided Souls so desperately wanting to be important men on Earth. And sadly, the only way they have found to do this is by hiding and destroying, by covertly shortening a person's lifespan, deteriorating the body, and mind control technology and hopefully breaking the spirit.

One thing is certain, we are dealing with evil in high places and an obvious powerful influence documented on this planet since the beginning of time.

Humanity consistently continues the search to uncover this particularly pernicious aspect of evil, revealed in the selected quotes below:

"The gods did smell the savour, the gods did smell the savour sweet, the gods gathered like flies around the man making sacrifice." – "The Epic of Gilgamesh"

"These Elementals live in the soul-realm of man as long as he lives, and grow strong and fat, for they live on his life-principle, and are fed by the substance of his thoughts."- Franz Hartmann, M.D.s, "Magic: White and Black"

"They took over because we are food for them, and they squeeze us mercilessly because we are their sustenance. Just as we rear chickens in chicken coops, gallineros, the predators rear us in human coops, humaneros. Therefore, their food is always available to them."- Carlos Castaneda, "The Active Side of Infinity"

"They are attracted to emotions. Animal fear is what attracts them the most; it releases the kind of energy that suits them."- Carlos Castaneda, "The Fire from Within"

"They were interested in emotion...They feasted as they made love to me."- Rick Strassman, M.D., quoting DMT volunteer speaking of insect-like, reptilian creatures in "DMT the Spirit Molecule"

The obvious hope through greater understanding is that by awareness of the undercurrent humanity can be separated from a cunning energetic evil, transformed from darkness to Light.

In reality, we are living in a superficial matrix of sorts whereas our thoughts can and are being used to create our reality, expertly explained in the Jean Baudrillard excerpt below entitled,

"Christ & Reality: Reading 'The Matrix' as a Postmodernist Statement"

"The concept of a "matrix" as an all-encompassing alternate reality is not new, nor did it originate with the 1999 film that is the subject of this essay. In his 1964 text entitled Understanding Media, Marshall McLuhan foreshadowed its coming, writing:

Literacy remains even now the base and model of all programs of industrial mechanization; but, at the same time, locks the minds and senses of its users in the mechanical and fragmentary matrix that is so necessary to the maintenance of mechanized society.

William Gibson's Neuromancer, generally acknowledged as the prototypical cyber-punk novel used the term to describe an artificial, alternate reality that he also referred to as "cyberspace:

The matrix has its roots in primitive arcade games...a consensual hallucination experienced daily by billions of legitimate operators, in every nation, by children being taught mathematical concepts... A graphic representation of data abstracted from the banks of every computer in the human system, of unthinkable complexity. (51)

Discussing the final passage into death, a 1994 translation of The Tibetan Book of the Dead says, "This last is followed by the consciousness taking up its abode in a suitable matrix, whence it is born again as a Birth-Consciousness." (ixix)

The film "The Matrix" borrows these concepts as its premise, but is far from a shallow exploitation of the idea. It is a testimony to postmodern thought; a monument erected for the sole purpose of deconstructing itself. The dystopia that is the film's setting problematizes the concepts of existence, reality, faith, perception and individual freedom. "The Matrix" is a multi- sensory exploration of the ideas of Jean Baudrillard, a deconstructive re-telling of the Christ-story, and an attack on the viewer's perception of reality. As Morpheus tells Neo, "The Matrix is the world that has been pulled over your eyes to blind you from the truth."

One thing is certain, I believe, when it is all said and done, at our time of departure from this Earthly experience, we will meet that which we resonated too and aligned our soul with by choice.

Our choices are darkness or light, good or evil, love or hate, negative or positive in the polar state of this existence as the evolutionary process continues.

This program particularly focuses on emotions of strong fear or hate. The control mechanism deliberately creates an environment on Earth that is as traumatized as possible and we are bombarded with negativity daily. This is for continued control of humanity that is being set-up to be at each other by creating conflicts, bloodshed, and raging wars.

Any person with low self-esteem is susceptible to attacks from the evil. When you don't think highly of yourself and the things you do, you are likely to let "somebody else" run things. If you don't control your mind, someone else will.

When a person is unbalanced in body and mind, when he or she hasn't dealt with all "the dark sides" of his or her being, courageously, then a "division" in the personality might take place, resulting in emotional turmoil. This is undesirable and dangerous because the person is only aware of one personality aspect at a time. Therefore, when the "dark personality" is active, or in the case of this program activated, the person can be entirely controlled.

However, we do not need any artificial tool to be in contact with your higher self or with the universal aspect of ourselves which is the connection with a higher power.

The fact is you are a perfect being with wonderful and powerful abilities. All you have to do is to revive your original force and let it flow. It involves letting go of that which has not served you in life, whether memories, thoughts, experiences, and by doing so consciously.

All the answers you need already exist in a cosmic memory bank, constituted as a field of subtle energies unlocking the door for balance and awakening.

In my case, realizing that those running this program, are factually trying to covertly murder, me, for nothing, the truth useless to them, in their possession, I continue to walk my path, through exposure.

If I die, sooner than the 90 years typical of my Family Tree and bloodline, healthy eating, exercising, etc., by some type of major beamed illness, lung, breast, or brain cancer etc., or even a focused beamed, zapped heart attack, as part of the high tech targeting, know that the beam continued to slow cook and irradiate my body each day by extremely low frequencies.

Let it be known it was the Los Angeles Police Department's corrupt secret police unit, and the Blue Code of Silence, with the operation spearheaded by the FBI who is factually America's secret police, now using beamed military technology and joined by military personnel from bases across the nation, in a joint high-tech COINTELPRO effort are the culprits, who believe above reproach.

Wake up to the understanding that this is simply the high-tech paradigm today in the long planned "Militarized Police State."

In this type of targeting, everything, and I mean everything, is well known about the target, to include dental and medical record information in the ever-present search for an edge, strategically, and apparently your life history, back to childhood as well.

Two things I noticed when reviewing the pulmonary results, after pneumonia, was that the CT scan did picked up Atherosclerosis and Pulmonary Arterial Hypertension with the later specifically during the time I had high-tech, believe it or not, pneumonia with this operation privy to this useful information as well, thus the lung cooking.

Because of this, I am yet again prompted to publicize how a person can be silenced bogusly under the guise of even high cholesterol, while monitoring the target's health status, although I fall into the low risk, non-threatening age group shown in the article below.

The question is...

"Is High Cholesterol Actually Bad for You?"

Dr. Sharon Orrange

Dr. Orrange is an Associate Professor of Clinical Medicine in the Division of Geriatric, Hospitalist and General Internal Medicine at the Keck School of Medicine of USC.

Posted on June 29, 2016

High cholesterol may be much ado about nothing, especially in older folks. A recent meta-analysis published in BMJ Open raises a strong argument that lowering LDL cholesterol in older people doesn't help at all.

Where does this leave us? Are we over-treating millions of folks with cholesterol lowering drugs, "statins" like Lipitor (atorvastatin), Crestor (rosuvastatin), and Zocor (simvastatin)? Let's take a look.

After reviewing 19 studies with over 68,000 people here are 6 things we know:

Previous research has shown a weak association between total cholesterol and death in folks over the age of 60. Most people in this current study were 65 – 85 years of age.

Many of you take cholesterol lowering drugs because some studies have shown you can lower your risk of death from stroke and heart disease by lowering LDL ("bad") cholesterol with statin medications. Turns out, that is seen most often in younger people.

This new widespread review showed no association between LDL cholesterol and mortality among people older than 60 years. So even high "bad" cholesterol was not found to increase risk of stroke and heart disease in people older than 60.

Our way of thinking about cholesterol may need to change. Because atherosclerosis (plaque buildup in your arteries) starts mainly in middle-aged people and becomes more pronounced with increasing age, there should be an even greater risk of heart attack over time—if untreated high cholesterol was a factor. Why is high cholesterol a risk factor for heart attack and stroke in the young and middle-aged, but not in elderly people? We don't really know.

What does this mean for the use of statins in older people? They are likely over prescribed. Most statin trials have shown little effect on cardiovascular disease and mortality, with best results showing a 2% reduction in mortality.

This review will again ignite the debate about the cause of atherosclerosis and heart disease and whether the benefits from statin treatment have been exaggerated.

Confused? Where are we now on the use of statins to lower cholesterol if we aren't even sure lowering LDL cholesterol in folks over 60 years of age helps? For now, those of you between the age of 40 and 65 should check out a risk calculator to see where you stand.

If your 10-year risk is higher than 7.5% according to the calculator, you should talk to your doctor about the risks and benefits of statins.

What do we know for sure? If you score above the 7.5 percent largely because you smoke or have high blood pressure, the best thing you can do is quit smoking or control your blood pressure. That may be enough to get you below the cutoff, so you don't need to start on statin therapy. Lifestyle changes also help reduce risk so diet and exercise changes are a no brainer. ~ Dr. O.

These technological technicians are well-aware what they can create using these bioelectric weapons, years in research and testing and development on many human guinea pigs.

Sadly, I must document that I join one in six today who have high cholesterol from eating at restaurants due to being single, divorced and an empty nest. This, lab information, however, impacted my immediately taking charge, and bringing my cholesterol down below the standard number and returning to a Vegetarian diet so that I would not become strategic prey and also give my body a fighting chance under the continue high-tech assaults.

My dad lived to be 90 years old with an enlarged heart.

If I die, the government killed me by approval of this insidious monstrosity passed down by approval from federal, to state and local police departments, as a covert, high-tech COINTELPRO targeting effort unifying military personnel and military technology around "lil ole me."

I join thousands targeted, and millions globally who understand that today technologically, if these operatives can't silence, destroy or entrap you, while trying to keep what they are really doing hidden, for DECADES, they apparently will kill or attempt too by slow deterioration of your health. And beamed Directed Energy Weapon, again, Breast Cancer reportedly is high on the list for women in opposition to this program and reporting it.

One thing I have learned, life is too short to not try and find every ounce of joy you can derive from it, no matter what. In fact, the exposure of this program, has given me great joy, empowering me, and the hope to save others my greatest reward.

The End or a New Beginning?

ABOUT THE AUTHOR

Researcher, Author, Human Rights Advocate, Blogger and Amazon Docudrama Creator related to Covert Human Experimentation.

Her professional website and blog are a gold mine of information, with thousands of views monthly focused on the covert use of psychophysical, psychological electronic frequency wave technology in full use today, as well as, the structure of an ongoing high-level nonconsensual human guinea pig program with combined technological harassment revealing corruption that makes other documented human experimentation historical programs look like child play. Her books, six in this series, this being the last, are not just detailed technical information but are an excellent resource for understanding the technology in use today, personal experiences, and the resulting beamed technological effects.

"The Program" is decades perfected and unleashed legally for testing, some would argue bogusly, for continued research activity, post 9/11, on men, women, and children, individuals, groups, communities and large populations.

It is absolutely no wonder there is an ongoing effort to discredit, silence, and subjugate Pittman. And, these are tactics confirmed by thousands of victims worldwide through the use of covertly beamed electromagnetic Directed Energy Weapon psycho-physical attacks designed to shut down Human Rights Advocates, bogusly labeled as

"Domestic Terrorist" by any means necessary and for using exposure to fight for their very lives.

www.ingramcontent.com/pod-product-compliance
Lightning Source LLC
Chambersburg PA
CBHW070859080526
44589CB00013B/1135